ENVIRONMENTAL REMEDIATION TECHNOLOGIES, REGULATIONS AND SAFETY

WILDFIRE SUPPRESSION

FEDERAL FUNDING AND SPENDING

ENVIRONMENTAL REMEDIATION TECHNOLOGIES, REGULATIONS AND SAFETY

Additional books in this series can be found on Nova's website under the Series tab.

Additional e-books in this series can be found on Nova's website under the e-book tab.

ENVIRONMENTAL REMEDIATION TECHNOLOGIES, REGULATIONS AND SAFETY

WILDFIRE SUPPRESSION

FEDERAL FUNDING AND SPENDING

NEAL JENKINS
EDITOR

nova publishers
New York

Library of Congress Cataloging-in-Publication Data

ISBN: 978-1-63483-773-6

Published by Nova Science Publishers, Inc. † New York

CONTENTS

PREFACE

The federal government is responsible for managing wildfires that begin on federal lands—such as national forests or national parks—while the states are responsible for managing wildfires that originate on all other lands. The federal government's wildfire management responsibilities—provided primarily through the Forest Service (FS) and Department of the Interior (DOI)— include prevention, detection, response, and recovery. The Federal Emergency Management Agency (FEMA) may also provide disaster relief for certain nonfederal wildfires. Congress appropriates wildfire management funding to both FS and DOI. Within their overall appropriations for wildfire, suppression operations are appropriated through two accounts for each agency: the Wildland Fire Management (WFM) accounts and the Federal Land Assistance, Management, and Enhancement Act (FLAME) reserve accounts. If the suppression funding in both of these accounts is exhausted during any given fiscal year, FS and DOI are authorized to transfer funds from their other accounts to pay for suppression activities. Congress also may provide additional funds for suppression activities through emergency or supplemental appropriations. Thus, for any given year, total suppression appropriations to FS or DOI may be a combination of three sources: the WFM suppression activity, the FLAME account, and supplemental appropriations, and the agencies also may access additional funding as needed through transfers. This book provides background information and analysis of funding for federal wildfire suppression operations; summarizes relevant legislative proposals introduced in the 114th Congress; discusses their possible implications; provides wildfire management appropriations data for FY2011-FY2015 and information on the President's FY2016 budget request for wildland fire management; provides general wildfire management statistics; and answers frequently asked questions about FMAGs.

In: Wildfire Suppression
Editor: Neal Jenkins
ISBN: 978-1-63483-773-6

Chapter 1

WILDFIRE SPENDING: BACKGROUND, ISSUES, AND LEGISLATION IN THE 114TH CONGRESS*

Katie Hoover[1], Bruce R. Lindsay[2], Francis X. McCarthy[3] and Jessica Tollestrup[4]

SUMMARY

The federal government is responsible for managing wildfires that begin on federal lands—such as national forests or national parks—while the states are responsible for managing wildfires that originate on all other lands. The federal government's wildfire management responsibilities— provided primarily through the Forest Service (FS) and Department of the Interior (DOI)— include prevention, detection, response, and recovery. The Federal Emergency Management Agency (FEMA) may also provide disaster relief for certain nonfederal wildfires.

Congress appropriates wildfire management funding to both FS and DOI. Within their overall appropriations for wildfire, suppression operations are appropriated through two accounts for each agency: the Wildland Fire Management (WFM) accounts and the Federal Land Assistance, Management,

* This is an edited, reformatted and augmented version of a Congressional Research Service publication R44082, prepared for Members and Committees of Congress dated July 7, 2015.

and Enhancement Act (FLAME) reserve accounts. If the suppression funding in both of these accounts is exhausted during any given fiscal year, FS and DOI are authorized to transfer funds from their other accounts to pay for suppression activities. Congress also may provide additional funds for suppression activities through emergency or supplemental appropriations. Thus, for any given year, total suppression appropriations to FS or DOI may be a combination of three sources: the WFM suppression activity, the FLAME account, and supplemental appropriations, and the agencies also may access additional funding as needed through transfers.

Congress is debating the level and direction of federal spending on wildland fire management. Wildfire spending has more than doubled since the 1990s, going from $1.6 billion in FY1994 to $3.9 billion in FY2014. A significant portion of that increase is related to rising suppression costs, even during years of relatively mild wildfire activity, although the costs vary annually and are difficult to predict in advance. Since FY2005, FS and DOI have required more suppression funds than had been appropriated to them seven times, leading the agencies to transfer funds from other accounts and prompting concerns that the increasing suppression spending may be coming at the expense of other agency programs. In response to each of these seven instances, Congress enacted supplemental appropriations to repay the transferred funds or to replenish the agency's suppression accounts. Further, wildfire spending—as well as all discretionary spending—is currently subject to certain procedural and budgetary controls. In the past, Congress has effectively waived some of these controls for certain wildfire spending, but it has not done so in more recent years. This has prompted some to explore providing certain wildfire spending effectively outside of those constraints.

To date, four bills have been introduced in the 114th Congress to address wildfire spending issues: H.R. 167, S. 235, S. 508, and H.R. 2647. All four bills, directly or indirectly, might allow for some wildfire suppression funds—subject to certain criteria—to be provided outside the statutory limits on discretionary spending, either through the annual appropriations process or through supplemental appropriations. Under these proposals, varying levels of wildfire funding would not need to compete with other programs and activities that are subject to the statutory limits. However, the amounts that could be provided for wildfire suppression operations under these proposals—both within and outside of the spending limits—would be subject to future appropriations decisions that could be made by Congress each fiscal year. These proposals also could affect certain funding mechanisms that have been used to provide additional spending for major disasters (e.g., hurricanes, earthquakes).

Federal funding for wildfire management—particularly for suppression operations on federal lands—raises several interrelated policy questions for Congress to consider. These questions include how much funding Congress should provide for suppression purposes—an activity whose costs are generally rising but vary annually and are difficult to predict. The federal agencies tasked with suppression activities may deplete their suppression resources rapidly, so Congress also may consider if, and how, to provide these agencies with quick access to additional funds to enable continued federal services in response to wildfires. In addition, Congress may address questions related to the source of the suppression funds, such as if rising suppression costs should be offset by cuts to other agency programs or if those costs should be considered outside of certain budgetary and procedural constraints. Further, Congress may consider options to enact various budgetary controls on suppression spending or other methods to constrain rising federal costs.

This report first provides background information and analysis of funding for federal wildfire suppression operations. The report concludes by summarizing relevant legislative proposals introduced in the 114th Congress and discussing their possible implications.

WILDFIRE BACKGROUND

The term *wildfire* is defined as an unplanned, unwanted wildland fire, including lightning-caused fires, unauthorized human-caused fires, and escaped prescribed fire projects.[1] States are responsible for responding to wildfires that begin on nonfederal (state, local, and private) lands, except for lands protected by the federal agencies under cooperative agreements. The federal government is responsible for responding to wildfires that begin on federal lands. The Forest Service (FS)—within the U.S. Department of Agriculture (USDA)—carries out wildfire management and response across the 193 million acres of the national forest system. The Department of the Interior (DOI) manages the wildfire response for more than 400 million acres of national parks, wildlife refuges and preserves, Indian reservations, and other public lands.[2]

The term *wildfire suppression* covers all of the work associated with extinguishing or confining a wildfire. Federal policy is generally to suppress wildfires unless a fire management plan identifies locations and conditions when monitoring or less suppression efforts are appropriate.[3] The primary federal responsibility for wildfire suppression is to protect lives, property, and resources on federal lands. The federal government has other wildland fire management responsibilities that include programs to prevent the future risk of catastrophic fires, such as by reducing the accumulation of hazardous fuels. The federal government also provides technical and financial assistance to states, local governments, and communities to protect nonfederal (both government and private) lands from wildfire damages. The federal government—primarily through the Federal Emergency Management Agency (FEMA)—also may provide disaster relief to state and local governments. In addition, FEMA may provide assistance to individuals and households if a major disaster declaration is issued as a result of the wildfire.[4]

Wildfire Statistics

Since 2000, an average of 74,000 wildfires burned an average 6.6 million acres every year (see **Figure 1**).[5] Over the past 10 years from 2005 to 2014, nearly 6.9 million acres burned annually on average. This figure is almost double the average annual acreage burned in the 1990s (3.6 million acres), although there were a greater number of fires annually (83,000 wildfires on average). The last two years have been below that average, with 3.6 million acres burned in 2014 and 4.3 million acres burned in 2013. These figures also were much smaller than the acreage burned in 2011 (8.7 million acres) and 2012 (9.3 million acres). Although fewer acres burned in 2014, there were more fires (63,606 wildfires) than in 2013 (47,579 wildfires). One percent of the fires in 2014 was classified as large or significant (666 wildfires), and 9 wildfires exceeded 40,000 acres in size.[6]

In 2014, 60% of the nationwide acreage burned by wildfires was on federal lands (2.1 million acres; see **Table 1**). The other 40% of the acreage burned occurred on state, local, or privately owned lands but also accounted for most of the fires (50,570 wildfires). This suggests that the fires that occur on federal land are larger in size, although fewer in number, compared with other landowners. This is particularly true in the West,[7] where less than half of the fires burned 71% of the acreage on federal lands compared to nonfederal lands. In the East, however, where there is less federal acreage, most of the fires and acreage burned occurred on nonfederal lands. Of the federal acreage burned nationwide in 2014, 40% burned on each of the FS and DOI's Bureau of Land Management (BLM) lands.

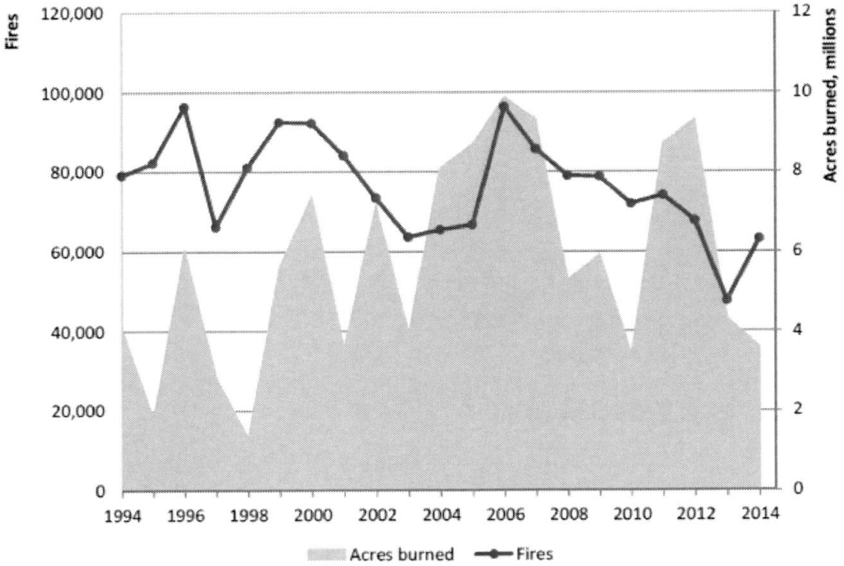

Source: CRS based on data from the National Interagency Fire Center (NIFC).

Notes: Data reflect wildland fires and acres burned nationwide, including wildland fires on federal and nonfederal lands.

Figure 1. Annual Trends in Wildfires and Acres Burned, 1994-2014.

Table 1. 2014 Wildfires and Acres Burned by Eastern and Western States

Land Ownership	Fires	Acres Burned
Western States	**20,712**	**2,736,684**
Federal	9,718	1,955,657
Forest Service	*5,732*	*824,986*
DOI	*3,856*	*1,101,550*
Other Federal	*130*	*29,121*
Nonfederal	10,994	781,027
Eastern States	**42,894**	**858,928**
Federal	3,318	197,359
Forest Service	*1,020*	*46,889*

Table 1. (Continued)

Land Ownership	Fires	Acres Burned
DOI	*2,202*	*139,797*
Other Federal	*96*	*10,673*
Nonfederal	39,576	661,569
TOTAL	**63,606**	**3,595,612**

Source: CRS. Data compiled from NIFC, State and Agency Fires and Acres reports, at http://www.nifc.gov/ fireInfo/fireInfo_statistics.html.

Notes: Western states are Alaska, Arizona, California, Colorado, Hawaii, Idaho, Montana, New Mexico, Nevada, Oregon, Utah, Washington, and Wyoming. Eastern states are all other states and include Puerto Rico.

Wildland-Urban Interface

Wildfires have been burning more land and threatening more structures in recent years. Nearly 2,000 structures were destroyed by wildfires in 2014. The area where structures (usually homes) are intermingled with—or adjacent to—vegetated wildlands (forests or rangelands) is called the wildland-urban interface (WUI).[8] More than one-third of all housing developments in the United States are located within the WUI.[9] In the West, nearly 900,000 homes are estimated to be at very high or high risk of wildfire damage in 2015.[10] While attention has focused on protecting life, property, and communities in the WUI, opinions vary over if and how much the federal government should pay to protect those resources.

U.S. federal wildland fire policy directs that the response to wildfire is to prioritize first the ecological, social, and legal consequences of the fire and then the economic costs of protection.[11] The policy further states that the cost of federal response to wildfire should be commensurate with the values to be protected (human, natural, historical, or cultural), but economic efficiency is not necessarily required. Response priorities include managing costs but without compromising safety.[12] While some believe this policy allows the federal land management agencies flexibility to provide a high-quality emergency response, others believe this is akin to a "blank check" policy and has removed any incentive for the agencies to control suppression costs.[13]

Federal wildfire suppression spending is influenced by several factors, including the size and intensity of the fire and the proximity of the fire to valuable resources (human, natural, historical, or cultural).[14] These resources require protection, which often increases firefighter risks as well as suppression costs as more personnel or assets are deployed to provide protection.[15] Federal suppression costs—daily, overall, and on a per-acre basis—become higher as the number and value of homes near a fire increase.[16]

When wildfire expenditures began to increase in the 2000s, many were concerned that the federal government was bearing too much of the cost of wildfires and that state, local, and private landowners lacked incentive to mitigate future fire risk to offset suppression costs.[17] The agencies have since modified their cost-share agreements with many of the states to provide more consistent arrangements, although these still may vary by state and by fire.[18] The agencies, particularly FS, also have initiated several technical and financial assistance programs to increase WUI community preparedness and homeowner protections. The 2006 USDA Office of Inspector General (OIG) report asked Congress to clarify the federal government's role in protecting WUI developments. However, the debate in more recent years seemingly has focused less on decreasing federal liability for rising suppression costs.

Federal Assistance for Nonfederal Wildfires

The federal government provides assistance for wildfires that begin on nonfederal lands. This assistance may come in several forms, including technical and financial assistance programs to mitigate the risk of future wildfire and direct response services under cooperative agreements. These cooperative fire protection agreements authorize federal and state partners to share resources—such as aviation equipment and personnel—depending on ongoing need during a wildfire season, allowing for a coordinated interagency response that deploys resources to areas of greatest critical need. The cost of these resources is then reimbursed as specified in the master agreement, which often lists several different methods to apportion costs, each with different financial impacts. This may include assigning the cost based on the proportion of acres burned within each agency's jurisdiction or on resources deployed, among others.[19] The National Interagency Coordination Center, located at the National Interagency Fire Center (NIFC), coordinates and allocates resources at a national level, and Geographic Area Coordination Centers coordinate and allocate resources at nine regional levels.[20]

Cooperative fire protection and financial and technical assistance programs are provided by the federal land management agencies—such as FS and DOI—but FEMA also may provide assistance to states, communities, and individuals during or after a wildfire.

Disaster Declarations

The term *fire* is included as an eligible event under the "Definitions" section of the Robert T. Stafford Disaster Relief and Emergency Assistance Act (Stafford Act).[21] A query of FEMA's declarations database for "wildfire" yields 27 major disaster declarations over a 60-year period (1953-2014).[22] That is an especially small subset given that there have been more than 2,200 major disaster declarations during that same period. This is partly because most wildfires are on federal lands, which make them ineligible for Stafford Act assistance. There are however, several types of declarations that provide assistance under the Stafford Act when the fires threaten state and private lands.

In the event that a wildfire is on state lands and there is a threat of several types of damage (e.g., to state and county infrastructure as well as private homes), a major disaster may be declared by the President if the governor of the affected state requests such assistance. Similarly, a governor could request an emergency declaration to address the threat before it warrants a major disaster declaration.[23] However, the most frequently employed authority for fighting wildfires under the Stafford Act is Section 420, which is specifically for "Fire Management Assistance."[24] This authority results more frequently in the grants that are discussed below.

Fire Management Assistance Grants

Section 420 of the Stafford Act authorizes the President to "declare" a Fire Management Assistance Grant (FMAG), which authorizes financial assistance to the requesting state.[25] A state must request an FMAG when the governor determines that a fire is burning out of control and threatens to become a major disaster. Typically, governors submit requests to the FEMA regional administrators. Requests can be submitted any time—day or night—and can be submitted verbally by telephone to expedite the process.

Once issued, the FMAG declaration authorizes various forms of federal assistance—such as equipment, personnel, and grants to state, local, and tribal governments—for the control, management, and mitigation of any fire on certain public or private forest land or grassland that might become a major disaster. The grants may reimburse up to 75% of the allowable suppression costs for eligible fires. It should be noted that FMAG declarations, unlike some major disaster declarations, do not authorize assistance to individuals and households.

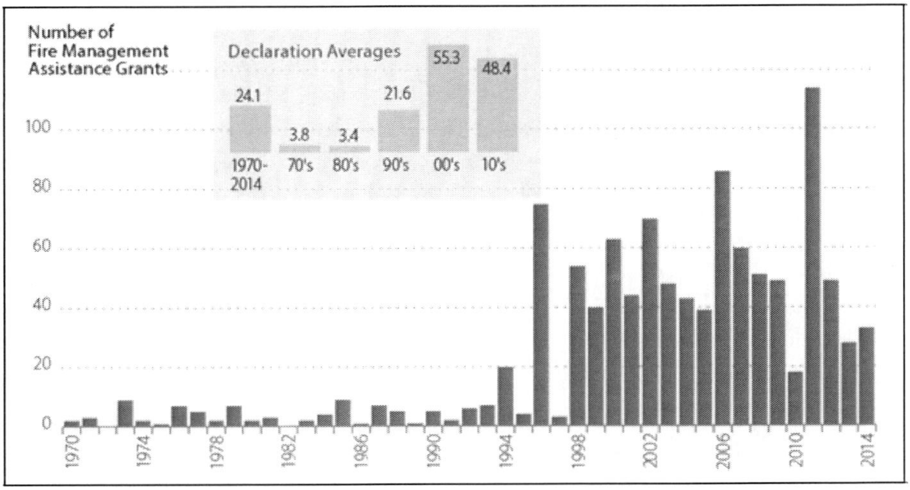

Source: CRS. Data provided by FEMA.

Figure 2. Fire Management Assistance Grants (1970-2014).

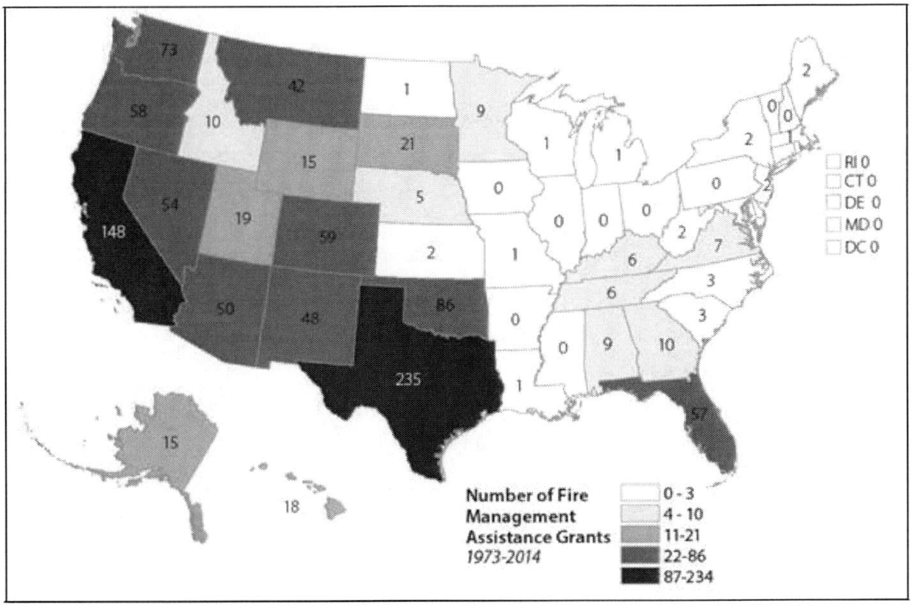

Source: CRS. Data provided by FEMA.

Figure 3. Fire Management Assistance Grants by State (1973-2014).

As shown in **Figure 2**, the number of FMAG declarations has increased in recent years, reaching a high of 114 in 2011. This surpassed the previous high of 86 FMAGs in 2006. As mentioned previously, FMAGS are designed to prevent fires from becoming major disasters. It could be argued that even though the cost for FMAG declarations may have increased, FMAGs may actually save federal dollars by reducing the need for a major disaster declaration, thus reducing overall spending on Stafford Act programs.

The first FMAG was declared in 1970, and they were rarely issued until the 1990s (see **Figure 2**). The average number of FMAGs declared in the 1970s and the 1980s was about three per year. During the 1990s, there were about 21 FMAG declarations per year (**Figure 3**). This upward trend continued into the 2000s, with an average of 53 FMAG declarations issued each year. Texas has received the most FMAGs (235 declarations), followed by California (148), Oklahoma (86), Washington (73), and Colorado (59) (see **Figure 3**).[26]

Congress has made efforts to expand the assistance authorized by FMAGs to include mitigation. For example, a provision in the FY2015 appropriations bill for the Department of Homeland Security (P.L. 114-4) allows FMAG funding to be used to mitigate the effects of future wildfire risk. The funding is calculated under Section 404, the Hazard Mitigation Grant Program (HMGP) of the Stafford Act.[27] The measure could allow the states receiving FMAGs in FY2015 to take mitigation efforts prior to the next fire threat (in addition to helping states control ongoing fires). The provision was included in the general provisions for P.L. 114-4, and therefore mitigation funding is available for FMAGs through the end of FY2015.[28] The HMGP program is funded by the Disaster Relief Fund (DRF) under major disaster declarations.

Disaster Relief Fund

Funds from the DRF, an account in the Department of Homeland Security Appropriations Act, are used to pay for ongoing recovery projects from disasters occurring in previous fiscal years, meet current emergency requirements, and serve as a reserve to pay for upcoming incidents. The DRF is funded annually and is a "no-year" account, meaning that unused funds from the previous fiscal year (if available) are carried over to the next fiscal year. In general, when the balance of the DRF becomes low, Congress provides additional funding through both annual and supplemental appropriations to replenish the account.[29]

In addition to major disasters, the DRF provides funding for emergency declarations and FMAGs as well as some administrative costs. This portion of the DRF is described in the Budget Control Act of 2011 (BCA; P.L. 112-25) as the "base/non-major disasters." The President's request for base/non-major disasters each fiscal year is based on a 10-year average for non-catastrophic events, whereas the request for major disasters is based on FEMA's spending plans for all past declared major disasters.

WILDFIRE MANAGEMENT APPROPRIATIONS

Federal funding for wildfire management is provided to both FS and DOI. Funding for DOI is provided to the department, which then allocates the funding to the Office of Wildland Fire and four agencies—the Bureau of Land Management (BLM), the Bureau of Indian Affairs, the National Park Service, and the U.S. Fish and Wildlife Service.[30]

Both DOI and FS receive annual discretionary appropriations through the Interior, Environment, and Related Agencies appropriations bills.[31] Each agency has two accounts for wildfire: a Wildland Fire Management (WFM) account and a Federal Land Assistance, Management, and Enhancement Act (FLAME)[32] account. Each agency's WFM appropriation is distributed among two subaccounts: *fire operations* and *other fire operations*. The *fire operations* subaccount receives the bulk of the WFM appropriation and funds two programs: preparedness and suppression. Appropriations for preparedness are used to support efforts that assist with fire prevention and detection, equipment, training, and baseline personnel. Suppression appropriations are primarily used for wildfire response. The *other fire operations* subaccount funds hazardous fuels reduction and fire assistance programs, as well as other activities that are more focused on decreasing the risk of future catastrophic wildfires.

Federal spending on wildfire management has more than doubled over the past two decades in terms of 2014 dollars (see **Figure 4**), although the trend has fluctuated annually. From FY1994 to FY1999, appropriations averaged $1.6 billion; since FY2000, appropriations have averaged $3.6 billion. In FY2008, total wildfire appropriations for FS and DOI were $4.9 billion in 2014 dollars ($4.5 billion in nominal dollars), the highest amount to date. Appropriations slowly declined in FY2012 to $3.2 billion (2014 dollars; $2.6 billion in nominal dollars), before increasing again in both FY2013 ($3.4 billion in 2014 dollars) and FY2014 ($3.9 billion). For FY2015, FS and DOI have received $3.5 billion in wildfire appropriations to date, but it is possible that this figure could increase before the end of the fiscal year if supplemental appropriations are enacted.

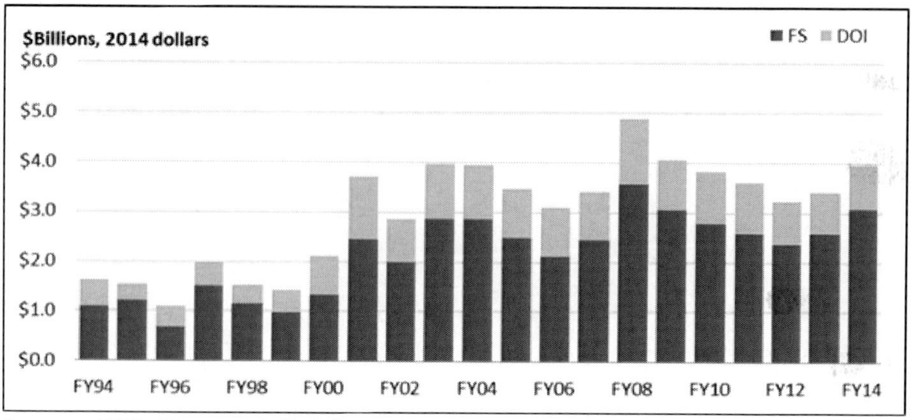

Source: CRS. Data compiled from detailed funding tables prepared by the House Committee on Appropriations.

Notes: Total wildfire management appropriations include appropriations to the WFM account, the FLAME reserve account, and any additional supplemental appropriations enacted for wildfire purposes to both FS and DOI. Figures adjusted to 2014 dollars using the annual GDP deflator price index reported by the U.S. Department of Commerce, Bureau of Economic Analysis, National Income and Products Accounts Tables, Table 1.1.9.

Figure 4. FS and DOI Total Wildfire Management Appropriations, FY1994-FY2014.

Table 2. FS and DOI Appropriations, FY2010-FY2015 (millions of nominal dollars)

	FY2010	FY2011	FY2012[a]	FY2013	FY2014	FY2015
Forest Service (FS)						
Total Suppression Appropriations	$1,410.5	$1,285.9	$853.6	$1,188.8	$1,595.5	$1,011.1
WFM Suppression	$997.5	$995.5	$538.2	$510.0	$680.5	$708.0
FLAME	$413.0	$290.4	$315.4	$299.0	$315.0	$303.1
Additional Appropriations[b]	NA	NA	NA	$380.0	$600.0	NA
WFM Other Than Suppression[c]	$1,181.2	$1,172.5	$1,436.6	$1,358.9	$1,481.8	$1,625.3

	FY2010	FY2011	FY2012[a]	FY2013	FY2014	FY2015
Total FS Wildfire Appropriations[d]	$2,591.7	$2,548.5	$2,290.2	$2,547.7	$3,077.3	$2,636.4
FS Appropriations Other Than Wildfire	$2,780.5	$2,626.6	$2,544.0	$2,377.0	$2,402.3	$2,420.0
Total FS Appropriations	$5,372.3	$5,085.0	$4,834.3	$4,924.7	$5,479.6	$5,056.2
% Wildfire Appropriations	48%	48%	47%	52%	56%	52%
Department of the Interior (DOI)						
Total Suppression Appropriations	$444.8	$459.8	$362.3	$375.9	$413.9	$383.7
WFM Suppression	$383.8	$399.0	$270.5	$261.2	$285.9	$291.7
FLAME	$61.0	$60.9	$91.9	$91.7	$92.0	$92.0
Additional Appropriations	NA	NA	NA	$23.0	$36.0	NA
WFM Other Than Suppression	$536.1	$519.1	$484.7	$426.3	$455.1	$513.1
Total DOI Wildfire Appropriations	$980.9	$978.9	$847.0	$802.2	$869.0	$896.8
DOI Approps. Other Than Wildfire	$10,066.3	$9,648.6	$9,452.8	$9,890.1	$9,605.5	$10,194.1
Total DOI Appropriations	$11,047.2	$10,627.5	$10,299.8	$10,692.3	$10,474.5	$11,090.9

Source: CRS. Data compiled from detailed funding tables prepared by the House Committee on Appropriations.

Notes: Totals may not add due to rounding. Data generally do not reflect any rescissions or budget adjustments for scorekeeping purposes.

a. Prior to FY2012, certain expenditures related to aviation assets and personnel costs were funded through the Suppression program; starting in FY2012, those costs were funded through the Preparedness program.

b. This includes any appropriations enacted for suppression purposes and titled as "supplemental," "additional," or "emergency" in the tables prepared by the House Committee on Appropriations.

c. This includes all appropriations to the FS WFM account, excluding funds appropriated to the Suppression program.

d. This includes all appropriations related to wildland fire management, including appropriations to the WFM account, the FLAME account, and any additional appropriations enacted for wildland fire management purposes.

The rising cost of wildfire management, combined with the annual spending fluctuations, makes budgeting for future wildfire spending difficult. Much of the increases, fluctuations, and unpredictability are driven by wildfire suppression costs. Analyzing wildfire cost trends is challenging, because the agency's account structures have changed; often the costs for one wildfire season (using a calendar year) are covered over two fiscal years, and sometimes appropriations are enacted in one fiscal year to cover costs incurred in previous fiscal years.

Suppression Appropriations

Suppression appropriations are used to control wildland fires on federal land, as well as wildland fires on nonfederal lands under fire protection agreements. Suppression operations fund firefighter salaries, aviation asset operations, incident support functions, and personnel and resources for post-wildfire response programs.[33]

Within the overall appropriations for wildfire, suppression operations are appropriated through two accounts for both FS and DOI: the suppression activity within the respective WFM accounts and the respective FLAME reserve accounts. These also are funded annually through the Interior Appropriations Act. If these suppression resources are exhausted during any given fiscal year, FS and DOI are authorized to transfer funds from their other accounts to pay for suppression activities. Congress also may fund suppression activities—including repaying borrowed funds from the previous fiscal year—through emergency or supplemental appropriations. (These processes and their impacts are discussed later in the report.) Thus, for any given year, appropriations to FS or DOI for suppression activities may be a combination of three sources: the WFM suppression account, the FLAME account, and supplemental appropriations (see **Table 2**); but the agencies also have access to additional funds through the transfer authority. See **Figure 5** and **Figure 6** for a breakdown of FS and DOI suppression appropriations, respectively.

Over the past five years, total FS suppression appropriations have averaged $1.2 billion per fiscal year, whereas total DOI suppression appropriations have averaged $399 million (see **Table 2**). Over the past 10 years, suppression activities, on average, have accounted for half of the FS's overall wildfire appropriation and a quarter of the agency's total appropriation (see **Figure 7**). Within DOI, wildfire appropriations—including suppression—are smaller and account for a significantly smaller portion of the overall DOI budget. Analyzing trends, however, is complicated because of certain structural changes FS and DOI have made to their wildfire accounts within the past five years. These changes include adding the FLAME account as well as moving certain aviation and personnel costs between the Suppression and Preparedness programs.

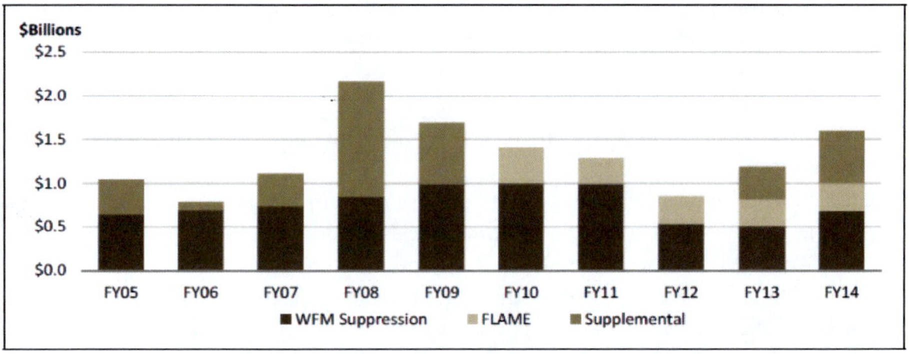

Source: CRS. Data compiled from detailed funding tables prepared by the House Committee on Appropriations.

Notes: The FLAME account was established in FY2010.

Figure 5. FS Suppression Appropriations, FY2005-FY2014 (dollars in billions).

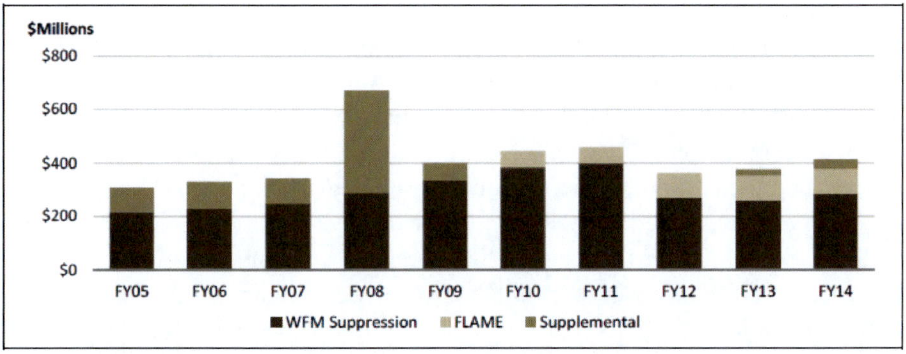

Source: CRS. Data compiled from detailed funding tables prepared by the House Committee on Appropriations.

Notes: The FLAME account was established in FY2010.

Figure 6. DOI Suppression Appropriations, FY2005-FY2014 (dollars in millions).

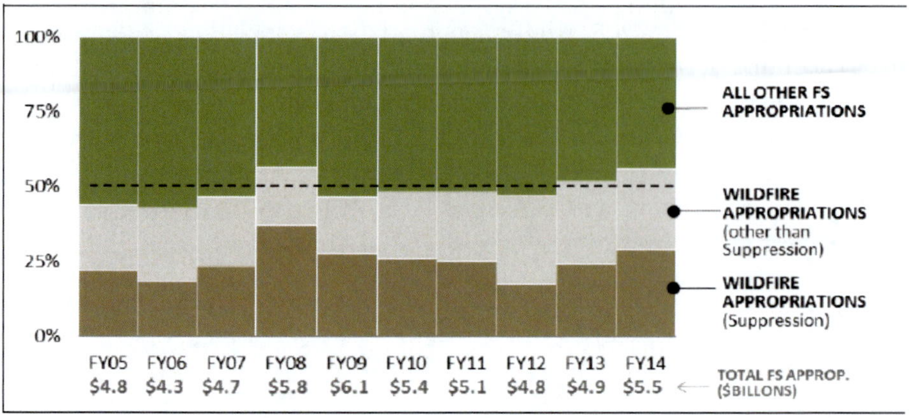

Source: CRS. Data compiled from detailed funding tables prepared by the House Committee on Appropriations.

Notes: "Wildfire appropriations (other than Suppression)" includes all appropriations to the FS WFM account, excluding funds appropriated to the Suppression program. "Wildfire Appropriations (Suppression)" includes funds appropriated to the WFM Suppression program, the FLAME reserve account, and any emergency or supplemental appropriations provided for suppression activities. "All Other FS Appropriations" includes appropriations to all FS accounts except the WFM account, FLAME reserve account, and any emergency or supplemental appropriations provided for suppression activities.

Figure 7. Distribution of FS Appropriations (FY2005-FY2014).

FLAME

Congress established a FLAME account—under the Federal Land Assistance, Management, and Enhancement Act of 2009[34]—for both FS and DOI in part to account for the growing cost of wildfire suppression.[35] The FLAME accounts provide a source of reserve funds used to cover the costs of large or complex fires or when amounts provided in their WFM suppression accounts are exhausted. Since the FLAME accounts were established in FY2010, the FS FLAME account has received an annual average of $323 million; the DOI FLAME account has received an annual average of $82 million (see **Table 2**). This represents about 30% of the combined (WFM and FLAME) appropriation for suppression activities for FS and 20% for DOI.

Both the Secretary of Agriculture and the Secretary of the Interior may transfer funds from their respective FLAME accounts into the respective WFM accounts for suppression activities upon a secretarial declaration.[36] The declaration may be issued if the fire covers at least 300 acres or threatens lives, property, or resources, among other criteria. Further, either Secretary may issue a declaration if his or her respective WFM suppression account is within 30 days of depletion. Any remaining FLAME funds may then be transferred into the WFM suppression account and used for wildfire response, regardless of the size or complexity of the fire. DOI reports that in 2014, 42 wildland fires were eligible for FLAME funding and $50 million was transferred from the DOI FLAME account.[37] FS did not declare individual fires for FLAME funding in 2014 but declared FLAME funds available for all fires due to the exhaustion of its WFM suppression funds.[38]

The FLAME Act also prohibited fire borrowing—transferring funds from other accounts to cover suppression obligations—unless and until the FLAME account is exhausted. Although FS reports that its FLAME account has been exhausted every year since FY2011—one year after the account was established—FS has transferred funds from other accounts only twice: in FY2012 and FY2013 (see **Table 3**).[39] DOI's FLAME account has been exhausted twice, in FY2012 and FY2013, and DOI transferred funds from other accounts in both of those years (see **Table 4**).[40]

Transfer Authority to Supplement Suppression Funds

During an active wildfire season, the agencies may deplete their suppression accounts quickly. However, they must continue to respond to wildfires and therefore need to be able to access additional funds in a timely manner. Therefore, Congress has granted FS and DOI the authority to transfer funds from other accounts and programs to ensure that federal emergency response activities continue under certain conditions (often referred to as *fire transfers* or *fire borrowing*). The transfer authority is granted annually in the Interior, Environment, and Related Agencies appropriations acts, specifically in the general provisions section for DOI and the administrative provisions section for FS. The authority to transfer funds for WFM-related activities was first granted in the FY1980 appropriations law (P.L. 96-126), which allowed transfers for the emergency rehabilitation of lands impacted by wildfire. The authority was continued annually, and then the FY1989 Interior appropriations law (P.L. 100-371) expanded the authority to allow for funds to be transferred for firefighting purposes in addition to emergency rehabilitation.[41] As noted above, the conditions for the transfer authority are that suppression funds in the respective WFM suppression account and FLAME reserve account must be nearly depleted.[42] Funds may be transferred from other discretionary accounts as well as from mandatory and permanent funding accounts and trust funds. Since the establishment of their FLAME accounts in FY2010, FS and

DOI each have borrowed from other accounts twice: in FY2012 and FY2013 (see **Table 3** and **Table 4**).

Table 3. FS Wildfire Suppression Spending, FY2004-FY2015 (millions of nominal dollars)

		Appropriations[b]				
Fiscal Year	Rolling 10-Year Suppression Obligation Average[a]	WFM Suppression and FLAME	Supplemental or Emergency Suppression	Total	Annual Suppression Obligations[c]	Funds Transferred from Other Accounts for Suppression
2004	$604.6	$597.1	$724.1	$1,321.2	$726.0	$0
2005	$685.4	$648.9	$395.5	$1,044.3	$690.0	$0
2006	$700.5	$690.2	$100.0	$790.2	$1,501.0	$200.0
2007	$746.2	$741.5	$370.0	$1,111.5	$1,374.0	$100.0
2008	$911.0	$845.6	$1,326.0	$2,171.6	$1,458.7	$273.8
2009	$993.9	$993.9	$700.0	$1,693.9	$1,018.3	$0
2010	$1,128.5	$1,410.5	$0	$1,410.5	$897.7	$0
2011	$886.0	$1,285.9	$0	$1,285.9	$1,414.4	$0
2012	$854.6	$853.6	$0	$853.6	$1,436.6	$440.0
2013	$931.0	$809.0	$379.9	$1,188.8	$1,356.5	$505.0
2014	$995.5	$995.5	$600.0	$1,595.5	$1,196.0	$0
2015	$1,011.6	$1,011.1	NA	$1,011.6	NA	NA

Source: Compiled by CRS. Unless otherwise specified below, data derived from detailed funding tables prepared by the House Committee on Appropriations, annual agency budget documents, and data from the FS legislative affairs office, July 2014 and updated January 2015.

a. Inflation adjusted for the fiscal year in which it is reported. This is the budget level requested by the President for Suppression (WFM Suppression and, starting in FY2010, FLAME).

b. Total Appropriations includes appropriations to FS's WFM suppression account, FLAME account, and any supplemental or emergency appropriation enacted for suppression activities, but it does not generally reflect any rescissions or budget adjustments for scorekeeping purposes. Emergency or supplemental appropriations may be used to repay funds borrowed from other accounts in the previous fiscal year.

c. Obligations may exceed appropriations in any given year because FS is authorized to carry forward unobligated balances from previous fiscal years and to transfer money from other accounts for suppression activities.

Table 4. DOI Wildfire Suppression Spending, FY2004-FY2015 (millions of nominal dollars)

		Appropriations[b]				
Fiscal Year	Rolling 10- Year Suppression Obligation Average[a]	WFM Suppression and FLAME	Supplemental or Emergency Suppression	Total	Annual Suppression Obligations[c]	Funds Transferred from Other Accounts for Suppression
2004	$195.3	$192.9	$234.4	$427.3	$281.2	$0
2005	$221.5	$218.4	$89.6	$308.0	$294.1	$0
2006	$234.2	$230.7	$100.0	$330.7	$424.1	$96.0
2007	$257.0	$249.2	$95.0	$344.2	$470.5	$112.5
2008	$294.4	$289.8	$384.0	$673.8	$392.8	$0
2009	$335.2	$335.2	$65.0	$400.2	$218.4	$0
2010	$369.8	$444.8	$0	$444.8	$231.2	$0
2011	$384.0	$459.8	$0	$459.8	$318.8	$0
2012	$362.6	$362.3	$0	$362.3	$465.8	$15.5
2013	$368.5	$352.9	$23.0	$375.9	$399.2	$34.1
2014	$377.9	$377.9	$36.0	$413.9	$326.2	$0
2015	$383.7	$383.7	NA	$383.7	NA	NA

Source: Compiled by CRS. Unless otherwise specified below, data derived from detailed funding tables prepared by the House Committee on Appropriations, annual agency budget documents, and data from the DOI Office of Wildland Fire Management, July 2014 and updated January 2015.

a. Inflation adjusted for the fiscal year in which it is reported.

b. Total Appropriations includes appropriations to DOI's WFM suppression account, FLAME account, and any supplemental or emergency appropriation enacted for suppression activities, but it does not generally reflect any rescissions or budget adjustments for scorekeeping purposes. Emergency or supplemental appropriations may be used to repay funds borrowed from other accounts in the previous fiscal year.

c. Obligations may exceed appropriations in any given year because DOI is authorized to carry forward unobligated balances from previous fiscal years and to transfer money from other accounts for suppression activities.

Typically, FS and DOI have developed an internal fire borrowing plan prior to the start of the wildfire season.[43] The plans identify accounts and programs that may be targeted if transfers are needed, based in part on unobligated balances and in part on an incremental strategy that depends on the amount that would need to be transferred while minimizing potential impacts to the public and agency programs.[44] Agencies often target programs that have relatively large unobligated balances.[45] These programs are often funded in one year, but the funds may not be obligated for several years, potentially allowing for transfers to be made with minimal immediate impact so long as the funds are reimbursed. The agencies may then also make a request to Congress to provide additional funding to replenish the FLAME accounts and to repay the transferred funds.[46]

Fire Borrowing Impacts

The authority to transfer funds from other agency accounts for suppression operations is controversial and has been since wildfire spending began to increase in the 2000s.[47] The authority to access additional funds for suppression operations provides FS and DOI flexibility to respond quickly in time-sensitive emergency situations. However, it also effectively provides them with an open-ended transfer authority, which some argue provides little incentive to manage suppression costs.[48] The agencies—and the Government Accountability Office (GAO)—also have argued that the fire transfers are disruptive to their other, non-fire operations and hinder their ability to carry out their statutory missions.[49] Borrowing from other program accounts—even when repaid in subsequent appropriations—creates uncertainty in the availability of funds and affects program implementation. In addition, some programs are time sensitive (e.g., land sales) and may suffer adverse impacts if and when delayed by fire transfers (e.g., changing land prices).[50]

Congress began introducing legislation to insulate agency appropriations from emergency fire suppression funding in the 110th Congress. The conferees of the FY2010 Interior appropriations bill stated their intent was that the funding provided in the FLAME account, together with appropriations to the WFM suppression account, should fully fund anticipated wildfire suppression needs and prevent future borrowing of funds from non-fire programs.[51] More recent legislative proposals also seek to prevent fire borrowing (see "Legislative Proposals in the 114th Congress" section of this report).

Supplemental Suppression Appropriations

When wildfire suppression funding is exhausted, Congress faces the question of reimbursing the accounts from which funding was transferred. The funds may be provided in an emergency appropriations bill (such as P.L. 108-324) or in the Interior appropriations bill but designated as "supplemental" or "additional." These funds may have been designated to repay fire transfers (usually made in a previous fiscal year) or to replenish the WFM suppression or FLAME accounts, although FS reports that $424 million in transferred funds remain unpaid since FY2002 (on a cumulative basis).[52] Due to the timing of the fire season (typically peaking in August), a reimbursement decision may be made after the end of the fiscal year when the transfers were made. This timing may complicate discussions about how much suppression funding is needed for the coming fiscal year.

Congress has provided supplemental appropriations in 7 of the 10 fiscal years since FY2005, funding $4.7 billion for emergency wildfire suppression activities for FS and DOI combined (see **Table 3** and **Table 4**). Since the establishment of the FLAME account in FY2010, Congress has provided supplemental suppression appropriations twice: $403 million in FY2013 and $636 million in FY2014. Supplemental appropriations provided in each of the years between FY2005 and FY2009 were designated so as not to be subject to certain procedural or statutory budget enforcement, such as the BCA limits on discretionary spending.[53] However, the supplemental appropriations provided in FY2013 and FY2014 did not contain any such designation and, as such, were subject to budget enforcement. (See the "Budgetary Constraints" section in this report.)

FORECASTING SUPPRESSION SPENDING

FS and DOI must estimate future suppression spending years in advance as well as during the wildfire season to forecast spending levels and account balances ongoing in the current fiscal year. The agencies formulate their budget requests for suppression operations using a rolling average of previous years' suppression spending (including supplemental appropriations enacted for suppression purposes).[54] This method originated in the 1990s from an agreement between the House and Senate Committees on the Budget, the Congressional Budget Office, and the Office of Management and Budget.[55] Prior to the enactment of the FLAME Act of 2009, the agencies' WFM suppression activity requests would equal their rolling 10-year suppression obligation averages. From FY2010 to FY2014, the agencies' WFM suppression activity requests plus their FLAME account requests equaled their rolling 10-year suppression obligation averages. Since FY2015, the Administration has requested a new funding mechanism for suppression operations, which includes eliminating the FLAME reserve fund and requesting 70% of the rolling 10-year suppression obligation average for the WFM suppression activity.[56]

Due to the timing of the budget process, the suppression budget request for any given year is based on the rolling 10-year suppression obligation average calculated two fiscal years previously. For example, the FY2016 suppression budget request was formulated using the FY2014 rolling obligation average. This means that suppression spending from FY2005 through FY2014 was used to formulate the suppression budget request for FY2016.

Because it is based on past spending, the rolling 10-year suppression obligation average is a lagging indicator of future suppression spending.[57] Lagging indicators, in general, demonstrate patterns across previous years but do not necessarily signal future trends. As such, the rolling 10- year suppression obligation average may not be the most accurate method to predict future suppression spending needs during the budget formulation process. For example, the rolling 10- year suppression obligation average has underestimated suppression spending 9 out of the past 11 years for FS (see **Table 3**); and 7 out of the past 11 years for DOI (see **Table 4**). On average, over the past 11 years, the rolling 10-year suppression obligation average has been 40% below the obligations for FS and 19% below the DOI suppression obligations.

When wildfire spending began to increase in the 2000s, GAO noted that the agencies' forecasting methods were insufficient, both in terms of annual and in-season budgeting.[58] FS reportedly analyzed alternative methods, but FS and DOI still use the 10-year suppression obligation average to formulate budget requests, even though suppression spending surpasses the estimate most years.[59] An earlier proposed version of the FLAME Act would have required the agencies to use a rolling 5-year suppression obligation average to formulate their budget requests.[60] A rolling 5- year average potentially would have predicted future suppression spending more accurately than a 10-year average, since the lower values from earlier years would drop out of the calculation. However, because a rolling 5-year average still would have been based on past spending, it also would have been a lagging indicator and likely would have underestimated suppression spending. The enacted version of the FLAME Act did not contain that provision, however. Instead, the FLAME Act requires the agencies to develop an estimate based on the best available science—the FLAME forecasts developed below—but does not direct that these be used to formulate budget requests.[61]

FS and DOI also must predict suppression spending during a wildfire season to ensure the availability of funds and to determine if and how much additional funds are going to be necessary. In response to FLAME Act requirements, FS and DOI began using regression models that incorporate weather and climate data, among other data, to forecast spending.[62] The models predict a range of suppression spending within a 90% confidence interval and are published four times per year: March, May, and June, with a September outlook for the upcoming year. The FLAME forecasts vary; for example, the estimate for the upper 90% confidence interval ranged from $1.570 billion for FS in the March 2014 forecast to $1.996 billion in the May 2014 forecast three months later. In the four years (2011-2014) that data are available, suppression obligations exceeded the FLAME median forecast for the FS three times and DOI four times. However, obligations exceeded the upper 90% confidence interval for both FS and DOI only once (see **Figure 8** and **Figure 9**).

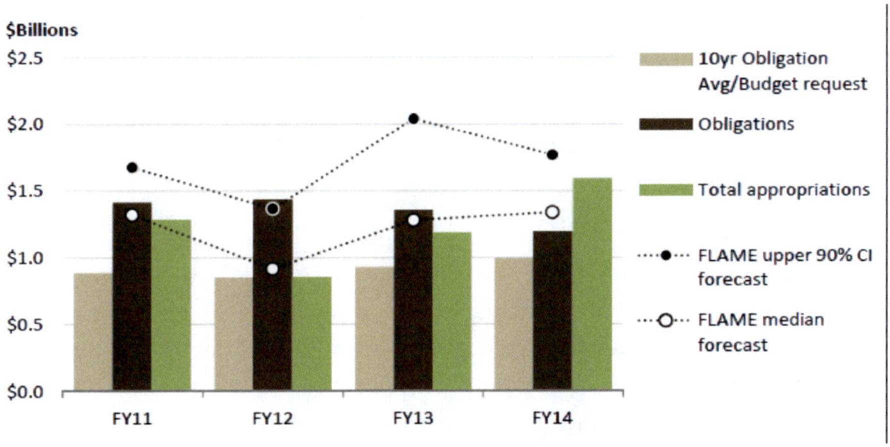

Source: CRS.

Notes: Suppression includes appropriations to the WFM suppression activity, FLAME account, and any supplemental or emergency appropriation enacted for suppression purposes. The FLAME median forecast is the average of the four reported median values per year. The FLAME upper 90% confidence interval forecast is the average of the four reported values per year.

Figure 8. FS Suppression Request, Appropriations, Obligations, and FLAME Forecasts (dollars in billions).

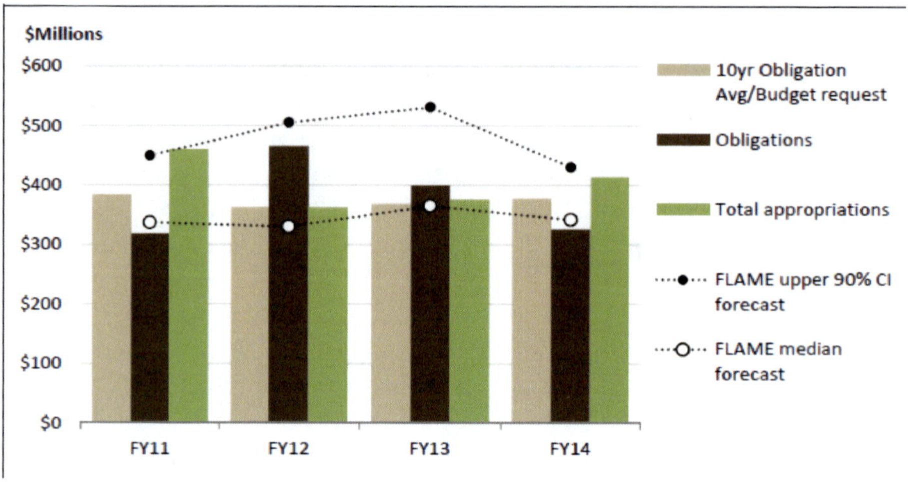

Source: CRS.

Notes: Suppression includes appropriations to the WFM suppression activity, FLAME account, and any supplemental or emergency appropriation enacted for suppression purposes. The FLAME median forecast is the average of the four reported median values per year. The FLAME upper 90% confidence interval forecast is the average of the four reported values per year.

Figure 9. DOI Suppression Request, Appropriations, Obligations, and FLAME Forecasts (dollars in millions).

The FLAME forecasts have not generally been used in the annual budget formulation process. However, the agencies use the forecasts to inform congressional decisionmakers about potential spending levels throughout a wildfire season or during budget hearings. In addition, the Administration's current budget request proposes providing access to funding up to the upper forecast range.

ISSUES FOR CONGRESS

Congress last addressed the mechanism for wildfire spending in the 110th Congress with the passage of the FLAME Act.[63] Several interrelated issues related to funding federal wildland fire management in general and wildfire suppression operations in particular have been topics of congressional hearings and proposed legislation in the 114th Congress.

Appropriation Levels and Forecasts

Each year, Congress considers at what level suppression (and wildland fire management in general) should be appropriated. Suppression costs are difficult to predict and can fluctuate widely. From FY2010 to FY2011, for example, combined FS and DOI suppression obligations increased more than 50%, but from FY2011 to FY2012, obligations increased by less than 10%. More recently, from FY2013 to FY2014, obligations decreased by 15%. These variations make it difficult for Congress to know at what level to appropriate in any given year. The budget formulation process is based on a rolling historic average, which has underestimated suppression spending 8 out of the last 10 years by nearly 50% annually on average. The agencies have had to borrow funds from other accounts in 5 of those years, and Congress has appropriated supplemental suppression appropriations in 7 of those years.

The intent of the FLAME accounts was to eliminate the need for fire borrowing or supplemental appropriations by serving as a reserve fund.[64] The FLAME accounts, however, are funded through a definite appropriation (a specified amount) that is still budgeted using the rolling 10- year suppression obligation average. After the FLAME accounts were established, the agencies continued to receive suppression appropriations equal to the 10-year suppression obligation average, although the funds were divided in two different accounts. The FLAME Act, in essence, created an additional account for suppression operations but did not create access to any additional funds above what was already being provided. The FLAME Act directed that the agencies develop a better formula to forecast suppression spending, but it did not mandate that the agencies use it to formulate their budget request (although that was debated). It also may be argued that the FLAME Act actually created various hurdles for the agencies to access a portion of the suppression funds they previously could freely access.

While it may be argued that the issue is that suppression costs are not being fully funded— requiring the agencies to deplete other accounts and, potentially, the appropriation of supplemental funds—the ability to access other funding sources has allowed the agencies to honor all suppression obligations incurred during any given fiscal year. However, this may sometimes be at the expense of not fully funding other programs. A more accurate description is that suppression costs for any given year are not necessarily being fully funded in advance for that fiscal year. In this sense, funding for suppression is sometimes reactive, not proactive. This is in part due to the unpredictable nature of wildfires and wildfire spending as well as difficulties in accurately predicting future suppression costs. Improving the suppression spending forecasts used for budget decisions may alleviate some of the agencies' needs to supplement their suppression accounts during the fiscal year. Congress may consider directing the agencies to use a different or specified methodology during their budget formulation processes, or it may continue to allow the agencies the discretion to formulate their own budget requests using the methodologies they deem most appropriate.

Funding Source

Federal spending on wildland fire management is currently more than double what it was in the 1990s. A significant portion of that increase is related to rising suppression costs, even during years of relatively mild wildfire activity. Although the 2013 and 2014 wildfire seasons had fewer fires and burned fewer acres than in 2012, Congress appropriated more funds in each successive fiscal year, including providing supplemental appropriations both years. However, on average, suppression appropriations have been slightly lower the last 5 fiscal years compared to the 10- year average (FY2005-FY2014: $1.9 billion for both FS and DOI; FY2010-FY2014: $1.7 average [2014 dollars]). Nonetheless, the overall trend is that costs are increasing relative to historical levels.

Congress may consider whether or how to address the rising costs of wildfire management and suppression operations. Some are concerned that the rising costs are coming at the expense of funding other agency programs, some of which may potentially reduce future suppression spending.[65] Wildfire accounts for an increasing proportion of FS's budget, up from 44% in FY2005 to 56% in FY2014. This means that other programs are receiving a smaller proportion of FS funds. For example, the proportion of FS's budget devoted to managing the national forest system decreased by 2% over that same time period, and the Construction Improvement and Maintenance proportion decreased by 6%, although it is not clear if the declines are attributable entirely to increasing wildfire funding. Congress may want to consider options that would maintain consistent levels of funding for other agency programs, although this could come at the expense of other agencies or programs funded through the Interior appropriations bill (discussed below) or may contribute to increasing the federal deficit if the costs are not offset.

Congress also may want to discuss ways to reduce the overall cost of suppression spending. FS, DOI, GAO, and others have argued that increasing current investments in hazardous fuel reduction projects may potentially reduce long-term suppression spending.[66] Others argue that the agencies should place a higher priority on cost containment efforts, perhaps by reducing agency budgets or restricting access to additional funds under certain conditions.[67] Another option Congress may consider is addressing the proportion of wildfire suppression costs borne by the federal government—for example, by developing a consistent federal cost apportionment method, which currently varies by state and often by fire, depending on the terms of the cooperative fire agreement. Other options include providing incentives for nonfederal entities to mitigate their risk of wildfire damages or discourage future WUI development, among others.

Budgetary Constraints

Discretionary spending—including wildfire appropriations—currently is subject to certain procedural and budgetary controls. In the past, Congress has effectively waived some of these controls for certain wildfire spending, but it has not done so in more recent years. This has prompted some to explore providing certain wildfire spending effectively outside of those constraints.

Discretionary Spending Limits Background

Pursuant to the Budget Control Act of 2011 (BCA),[68] discretionary spending currently is subject to statutory limits for each of the fiscal years between FY2012 and FY2021. Specifically, two different limits apply—one to "defense" and the other to "nondefense" spending. The defense category includes all discretionary spending under budget function 050 (defense); the nondefense category includes discretionary spending in all other budget functions.[69] Enacted discretionary spending may not exceed these limits. In the event that spending is enacted that exceeds a limit, the limit is to be enforced through sequestration, which involves the automatic cancellation of budget authority through largely across-the-board reductions of nonexempt programs and activities.[70] These reductions would occur to spending in the category of the limit that was breached.

Certain spending is effectively exempt from the discretionary spending limits pursuant to Section 251(b) of the Balanced Budget and Emergency Deficit Control Act (BBEDCA),[71] because those limits are "adjusted" upward to accommodate that spending. For example, these adjustments include budget authority designated as emergency requirements, as well as that designated as for disaster relief. The purpose of these designations is, in part, to address the unpredictable nature of disaster assistance while attempting to impose discipline on the amount spent by the federal government on disasters. The emergency designation could apply to a number of different types of purposes,[72] whereas the disaster relief designation is more narrowly limited to costs of major disasters that have been declared pursuant to Section 102(2) of the Stafford Act.[73] Wildfire suppression operations are an allowable purpose for the emergency designation but are not included under the disaster relief designation. Of the seven supplemental appropriations provided for wildfire suppression operations from FY2005 through FY2015, Congress designated five as emergency spending. The two most recent supplemental appropriations (FY2013 and FY2014) did not receive emergency designations.

While there is no limit on the amount of budget authority that can be designated as emergency requirements each fiscal year, the amount that can be designated using the disaster relief designation is limited by a formula. This formula is based on the average funding provided for disaster relief over the last 10 years, excluding the highest and lowest annual amounts.[74] If less than the maximum amount allowed by the formula is ultimately appropriated for a fiscal year, then the difference is added to the amount that the disaster formula yields for the next fiscal year.[75] For example, if the formula for a fiscal year allows $10 billion in disaster designated spending, but only $6 billion is appropriated, then the amount of the allowable disaster designation for the following fiscal year will be the formula plus the $4 billion in carryover from the previous fiscal year. This total amount—formula plus carryover, if any—is sometimes referred to as the *disaster cap*.

Since FY2012, when the BCA discretionary spending limits were first implemented, three different types of discretionary spending have been provided to prevent or respond to natural disasters—spending that is subject to the discretionary spending limits, spending designated as for disaster relief, and spending designated as for emergency requirements.

- Spending subject to the limits has been provided each fiscal year in regular appropriations measures to respond to disaster events, including wildfires, in a number of different appropriations accounts, such as certain FEMA appropriations in the Disaster Relief Fund[76] and the Small Business Administration appropriations for disaster loans.
- Disaster-designated spending has been provided each fiscal year through the regular appropriations process for response and recovery work for presidentially declared events that are not wildfires. In

previous practice, the great majority of such spending also has been appropriated to the FEMA DRF to be used for the response to major disasters. However, in some instances, disaster-designated funding also has been provided to address the unmet needs created by major disasters beyond those provided to the DRF, such as for the Community Development Block Grant program at the Department of Housing and Urban Development and disaster-related programs of the Army Corps of Engineers.

- Emergency spending usually has been provided to supplement disaster- designated spending, as well as spending that is subject to the discretionary spending limits, in response to events that have already occurred, including wildfires.

According to FEMA, the BCA necessitated the development of a new, two-part approach to accounting for disaster-related activity, with one approach for major disasters and another for all other DRF activity:

> Essentially, requests for DRF funding for FEMA's Stafford Act programs and disaster support activities fall into two categories: disaster relief cap adjustment and base/non-major disasters. Funding requested under the disaster relief cap adjustment is for major disasters declared pursuant to the Stafford Act and designated by the Congress as being for disaster relief pursuant to Section 251(b)(2)(D) of the BBEDCA, as amended by the BCA. Funding requested under the base/non-major disasters category includes Emergencies, Pre-disaster Surge Support, Fire Management Assistance Grants and activities that are non-disaster specific, such as Disaster Readiness Support (DRS) activities (e.g., distribution centers, reservist training, etc.).[77]

LEGISLATIVE PROPOSALS IN THE 114TH CONGRESS

The 114th Congress is considering options for addressing wildfire suppression spending issues, which include

- what level to appropriate for suppression activities;
- how to provide for unpredictable costs quickly;
- where the money should come from; and
- whether suppression money should be subject to or outside of certain statutory budget controls.

This section of the report discusses four proposals—S. 235, H.R. 167, S. 508, H.R. 2647—that have been introduced to address issues related to wildfire funding.

All four bills contain provisions that are intended to address concerns that wildfire suppression activities have not been adequately funded each year. Three of the bills—S. 235, H.R. 167, and S. 508—effectively would allow some funds provided for wildfire suppression activities that meet certain criteria to not be subject to the limits on discretionary spending (see **Table 5**). Under these proposals, varying levels of wildfire funding would not need to compete with other programs and activities that are subject to the limits. These proposals also would modify the existing disaster relief designation to interact with the additional wildfire spending that was provided pursuant to these new procedures. H.R. 2647 potentially could provide a related discretionary spending option for wildfire suppression operations funding, but it takes a different approach from the other three proposals.

S. 235 and H.R. 167 (The Wildfire Disaster Funding Act)

Two similar versions of the Wildfire Disaster Funding Act have been introduced in the 114th Congress (H.R. 167 and S. 235). The bills have minor differences from each other and from versions introduced in the 113th Congress (S. 1875 and H.R. 3992). S. 1875 was referred to the Senate Budget Committee; H.R. 167 was referred to the House Budget Committee, in addition to the Agriculture and Natural Resources Committees. This section of the report provides an analysis of these 114th Congress proposals as introduced.[78]

Summary

235 and H.R. 167 would amend the BBEDCA to add a new adjustment to the nondefense discretionary spending limit for "wildfire suppression operations," which includes spending for the purposes of

- the emergency and unpredictable aspects of wildland firefighting, including support, response, and emergency stabilization activities;
- other emergency management activities; and
- funds necessary to repay any transfers needed for these costs.

Table 5. Comparison of Selected Attributes of H.R. 167, S. 235, and S. 508

Select Provisions	H.R. 167	S. 235	S. 508
Adds adjustment for wildfire spending that is effectively not subject to statutory discretionary limits	Yes	Yes	Yes
Precondition(s) for adjustment	Yes	Yes	Yes
	70% 10-yr suppression obligation average must be appropriated within discretionary statutory limits annually	70% 10-yr suppression obligation average must be appropriated within discretionary statutory limits annually	The maximum of: • 100% 5-yr suppression obligation average; or • FLAME upper 90% CI must be appropriated within discretionary statutory limits annually An amount equal to at least 50% of the above amount must also be appropriated for specified forest management activities Funds in FLAME account must be expended

Table 5. (Continued)

Select Provisions	H.R. 167	S. 235	S. 508
Maximum adjustment	$2.689 billion/FY through FY2023	Increases annually: $1.410 billion FY2016 to $2.690 in FY2025	$1 billion/FY through FY2022
Interacts with disaster cap adjustment/DRF formula	Yes	Yes	Yes
Suppression Forecasting	Would require that the 10-yr suppression obligation average continue to be used for suppression budgeting	Would require that the 10-yr suppression obligation average continue to be used for suppression budgeting	Would require that the 5-yr suppression obligation average be used for suppression budgeting
Fire Transfers	Not addressed	Not addressed	Prohibited

Source: CRS.

There is a precondition for using this adjustment—that 70% of the 10-year average funding level for wildfire suppression operations must be appropriated subject to the limits. Once that precondition is met, the amount of the adjustment for the additional wildfire suppression operations funding is capped each fiscal year. In H.R. 167, the wildfire adjustment is limited to $2.689 billion each fiscal year through FY2022. In contrast, the maximum amount of the wildfire adjustment in S. 235 starts at $1.410 billion in FY2016 and gradually increases to $2.690 billion in FY2025.[79]

This proposal also affects the calculation of the maximum amount for the existing disaster relief adjustment—also referred to as the disaster cap—because the disaster cap calculation interacts with the proposed wildfire adjustment in two ways. First, starting in FY2017 under H.R. 167 or FY2018 under S. 235, the calculation of the 10-year rolling average for the disaster cap formula would include any budget authority provided using the wildfire designation in previous fiscal years. As a consequence, the amounts that are provided under the wildfire adjustment would gradually be incorporated into the calculation of the 10-year rolling average for disaster relief. Second, after the disaster cap for a fiscal year is calculated (including incorporating the amount of the prior-year wildfire adjustments into the calculation), the amount of the disaster cap is to be reduced by the amount of the previous fiscal year's wildfire adjustment. For example, if $1.4 billion was provided using the wildfire adjustment in FY2016, the FY2017 disaster cap would be lowered by $1.4 billion.

Potential Implications

Both S. 235 and H.R. 167 would remove some budget process barriers for the provision of additional wildfire suppression funds, either through the annual appropriations process or through supplemental appropriations. Whatever amount, if any, Congress elects to appropriate over the minimum 70% for wildfire suppression (up to the specified maximum in each bill) effectively would not be subject to the discretionary spending limits established in the BCA each year. For example, were these proposals to be in effect for FY2016, Congress could appropriate the minimum of $1.057 billion for suppression operations, which is 70% of the combined FS and DOI suppression obligation for FY2016 ($1.510 billion). But then Congress could appropriate an additional $2.689 billion under H.R. 167 or $1.410 billion under S. 235 in FY2016, effectively outside of the budget controls discussed above. This means the agencies could be appropriated up to $3.746 billion in H.R. 167, or $2.467 billion in S. 235, with $1.057 billion of those amounts being subject to the BCA discretionary limits. However, because these proposals leave actual funding decisions to future Congresses, Congress may choose to appropriate less than the minimum 70% threshold so that the adjustments allowed would not be triggered.

Other potential budgetary and policy effects of both S. 235 and H.R. 167 are unknown. It is not clear how these proposals would interact with the FLAME accounts. The bills specify that at least 70% of the budget request would have to be appropriated for suppression operations but do not specify whether the appropriations have to be in the agencies respective WFM or FLAME accounts. The FLAME accounts will expire if funds are not appropriated to (or withdrawn from) them for three consecutive years.[80]

S. 235 and H.R. 167 also are silent as to fire borrowing. If either of these proposals were to be enacted, it is unclear if fire borrowing authorities would continue to be provided in appropriations laws.

Both S. 235 and H.R. 167 would require the USDA and DOI Secretaries to continue to use the rolling 10-year average of suppression obligations to formulate their suppression budget requests. The USDA and DOI Secretaries also are directed to "promptly" submit any request to Congress for supplemental wildfire suppression operations funding, and they must submit a plan to Congress explaining how the supplemental appropriations will be obligated within 30 days of receiving them. This proposal would largely codify current agency practices, with some modifications. This may give Congress more advance notice about the need for supplemental wildfire suppression operations appropriations, as well as information on how the agencies propose to use those additional appropriations. Both this advance notice and additional information may aid Congress in appropriations decisionmaking and oversight. They also might add to the workload of both the FS and DOI during a critical time of providing wildfire assistance.

This interaction between the proposed wildfire and existing disaster relief adjustments has a number of potential implications for the amount of the disaster cap in the future. However, because these potential implications are highly sensitive to future appropriations decisions that will be made by Congress and the President each fiscal year, the extent to which the calculation for the disaster cap will be affected by the use of the wildfire adjustment cannot be determined definitively. Assuming that the wildfire adjustment is first used in FY2016, for the next several years, the disaster formula would be based on 10 fiscal years of funding for disaster relief (minus the highest and lowest fiscal year), but fewer than 10 fiscal years of the wildfire adjustment would be available to be incorporated into the base calculation. This has the potential to make the amount available for the disaster cap lower than it otherwise might have been. On the other hand, appropriations for the full adjustment allowed by the disaster cap have not been made since FY2013, which is causing the amounts that are used to calculate the 10-year rolling average for the amount of the cap in future fiscal years to be lower than they would have been had the full amount of the disaster cap been appropriated. As a consequence, adding wildfire-designated spending into the formula might have the effect of increasing the amount of the disaster cap above what it otherwise might have been. Another factor that affects these considerations is the amount that is assumed to be appropriated using the wildfire adjustment each fiscal year because either the full amount allowed, or a lesser amount might be appropriated.

Legislative Action

Neither of these proposals has received committee consideration as of the date of this report, although other relevant action has occurred in the context of the wildfire management budget request, the FY2016 budget resolution, and the FY2016 Interior Appropriations bill.

The Administration's FY2016 wildfire management budget request included a proposal to change how suppression operations are funded, similar to H.R. 167 and S. 235. However, the Administration's proposed maximum cap adjustment would be the difference between the upper 90% confidence interval FLAME forecast and the rolling 10-year suppression obligation average. (In FY2016, this would be $854.6 million.)[81]

The FY2016 budget resolution (S.Con.Res. 11) includes two procedural provisions related to wildfire suppression operations funding. First, the Senate Budget Committee-reported version included a spending-neutral reserve fund to "improve forest health" that, in part, would allow for the applicable committee allocations, aggregates, and other levels in the budget resolution to be revised to the purposes of measures relating to "reform of the process of budgeting for wildfire suppression operations."[82] Second, during Senate floor consideration of the budget resolution, an amendment was adopted that would provide a procedural adjustment to the budget enforcement limits associated with the budget resolution "if a measure becomes law that amends the adjustments to discretionary spending limits established under Section 251(b) of the Balanced Budget and Emergency Deficit Control Act of 1985 (2 U.S.C. 901(b)) for wildfire suppression funding" (S.Amdt. 434).[83] This amendment was agreed to by unanimous consent. Both of these provisions, with minor changes, were included in the final version of the budget resolution that was agreed to by Congress—at Sections 3208 and 4319.

The FY2016 Interior appropriations bill (S. 1645) as reported by the Senate Appropriations Committee also would address wildfire funding in a manner similar to H.R. 167 and S. 235. Title V of this bill would create a new adjustment for wildfire suppression operations and alter the calculation for the disaster cap based on the amount of wildfire adjustment appropriations that are ultimately provided. There are three main differences between the proposals discussed in this subsection and Title V of the Interior bill:

1. The precondition for using the adjustment is that 100% of the 10-year average for wildfire suppression operations also must be appropriated subject to the BBEDCA spending limits.
2. The range that is specified for the wildfire adjustment is set initially at $1.460 billion for FY2016 and gradually increases to $2.650 billion for FY2021.
3. FS and DOI would be required to submit a comprehensive report analyzing management decisions, fire activity, and fire accounting, among other topics.

S. 508 (The FLAME Act Amendments of 2015)

The FLAME Act Amendments of 2015 was introduced in the 114th Congress as S. 508 and referred to the Senate Committee on Energy and Natural Resources. An earlier version of the bill was previously introduced in the 113th Congress (S. 2593) but received no congressional action. This section of the report provides an analysis of the wildfire spending provisions of S. 508 as introduced. (S. 508 also contains additional titles addressing specific national forest management issues as well as stewardship contracting, which are not discussed.)

Summary

Like the proposals discussed in the "S. 235 and H.R. 167 (The Wildfire Disaster Funding Act)" section of this report, S. 508 would amend the BBEDCA to add a new adjustment for "wildfire suppression operations" spending. However, this proposal would impose two different sets of preconditions to access the adjustment. First, a minimum amount that is subject to the statutory discretionary spending limits must be appropriated to the DOI and FS Wildland Fire Management accounts. This minimum amount is the greater of

- 100% of the average costs for wildfire suppression operations over the previous five years; or
- the estimated amount of anticipated wildfire suppression costs at the upper bound of the 90% confidence interval for that fiscal year calculated in accordance with the FLAME Act.

In addition, an amount equal to at least 50% of that minimum amount must be specified for various forest management activities to mitigate future fire risk, such as

- authorized hazardous fuels reduction projects[84] and other activities of the Secretary of the Interior;
- forest restoration and fuel reduction activities performed outside of the WUI that are on condition class 3 federal land or condition class 2 federal land located within fire regime I, fire regime II, or fire regime III;[85] and
- timber sales, pre-commercial thinning, and salvage harvests performed on National Forest System lands.

Second, S. 508 also requires that all amounts in the FLAME fund have been expended as of the day before the date of enactment of an appropriation that would be subject to the wildfire adjustment.

Once those preconditions have been met, the measure would provide an adjustment for wildfire suppression operations of up to $1 billion in additional new budget authority in each of FY2016 through FY2022. The allowable purposes for spending that uses this adjustment are the same as those in S. 235 and H.R. 167.

This proposal also contains provisions that would affect the wildfire adjustment with the existing disaster designation but in a different way than the proposals discussed in the previous section. The formula for the disaster cap would still involve the 10-year rolling average for disaster relief spending, minus the high and low fiscal years, plus any prior year carryover. The amount of the wildfire adjustment from the previous fiscal year is also to be subtracted from that calculation. However, the 10-year rolling average for the disaster relief designation is not to incorporate amounts appropriated pursuant to the wildfire adjustment.

Potential Implications

The framework of S. 508 is similar to H.R. 167 and S. 235 in that it also would remove some budget process barriers for the provision of additional wildfire suppression funds. However, S. 508 differs from those other proposals in some significant ways. First, under S. 508 only up to an additional $1 billion for suppression spending could be provided under the new adjustment. Second, the minimum amount that would be required to be appropriated subject to those limits would be the greater of the rolling five-year suppression obligation average or the upper bound of the FLAME forecast. In addition, whatever amount Congress appropriates for suppression operations, an amount equal to half would be required for various forest management activities, such as hazardous fuels reduction. It is not clear if this other amount would be half of the suppression operations appropriation or an additional amount equal to that half. Regardless, requiring an appropriation amount equal to half of the greater of the five-year suppression obligation average or upper 90% FLAME forecast would increase the agencies' investment in hazardous fuels reduction substantially. Over the past five fiscal years, an average of $327 million has been provided to the FS, and $163 million to DOI, for the hazardous fuels activity within the agencies' respective WFM other operations accounts. While this may fund activities that would potentially mitigate future fire risk, these additional funds might come at the expense of other agency programs.

Another difference between the proposals discussed in the previous section and S. 508 is that it would require that all amounts in the FLAME fund be expended as of the day before the date of enactment of an appropriation that would be provided under the wildfire adjustment. This requirement has some potential implications for the timing of those additional wildfire appropriations. If the FLAME fund continued to receive annual appropriations, the wildfire adjustment generally could not be used to provide additional funds in the same annual appropriations laws. Instead, such funds would need to be provided in supplemental appropriations acts once the FLAME fund was exhausted each year, which could further add to Congress's summer legislative agenda. Alternatively, if Congress wants to provide these additional wildfire appropriations through the annual appropriations process, one way to do so would be to entirely discontinue providing appropriations to the FLAME fund and instead fund agency wildfire suppression operations entirely through their other accounts.

Like H.R. 167 and S. 235, the S. 508 formula also has potential implications for the amount that is available for the disaster cap in future fiscal years. Because the amount that is appropriated using the wildfire adjustment is subtracted from the disaster calculation and is not factored into the 10-year rolling average, it appears that the use of the wildfire adjustment has the potential to make the disaster cap lower than it otherwise might have been. However, the extent to which this would be the case would depend on whether the full amount allowed for the wildfire adjustment or a lesser amount was appropriated. As was the case for S. 235 and H.R. 167, these considerations are highly sensitive to future appropriations decisions that will be made by Congress and the President each fiscal year. Consequently, the extent to which the calculation for the disaster cap will be affected by the use of the wildfire adjustment cannot be definitively determined.

S. 508 also amends the FLAME Act. The agencies would continue to have their FLAME accounts for the most severe, complex, and costly wildfires, but the declaration criteria to access those funds would be changed: the minimum fire size would be 1,000 acres (up from 300 acres), or if the fire was within 10 miles of an urbanized area, but the Secretaries would still be able to access FLAME funds if their respective WFM suppression accounts were about to be depleted. S. 508 would also prohibit the agencies from transferring funds from other accounts for suppression purposes.

S. 508 is the only bill to address suppression forecasts. S. 508 would require FS and DOI to calculate and report the rolling five-year suppression obligation average and the outyear FLAME forecasts in their annual budget requests. The other bills continue the rolling 10-year suppression obligation average.

Legislative Action

S. 508 has not received any congressional action as of the date of this report. See the S. 235 and H.R. 167 "Legislative Action" subsection above for a discussion of action related to the wildfire management budget request, the FY2016 budget resolution, and FY2016 Interior appropriations.

H.R. 2647 (The Resilient Federal Forests Act of 2015)

The Resilient Federal Forests Act of 2015 was introduced in the 114th Congress as H.R. 2647 and referred to the House Committees on Agriculture and Natural Resources. This section of the report provides an analysis of the wildfire suppression provisions contained in Title IX of the measure, as included in the House Committee on Rules print dated July 1, 2015.[86] (H.R.

2647 also contains additional titles addressing other federal forestry issues, which are not discussed in this report.)

Summary

Title IX of H.R. 2647 would broaden the purposes for major disasters under the Stafford Act to include "wildfires on federal lands" (§901),[87] and it would create a related budgetary mechanism that potentially could be used to fund the response to each wildfire on federal land that was declared to be a major disaster (§902). Under this proposal, the affected federal land management agency (either the DOI Secretary or the USDA Secretary) may request a presidential declaration of a major disaster for wildfire on federal lands if the following requirements are satisfied:

1. The Secretary must make the request in writing.
2. The Secretary must certify that wildfire suppression operations appropriations that have been enacted for that fiscal year equal or exceed a certain minimum level, which is the rolling 10-fiscal year average cost of wildfire suppression operations incurred by the federal land management agencies. This minimum level includes any concurrently enacted rescissions. The unobligated balance of wildfire suppression operations funding, presumably from previous fiscal years, also may be credited to that minimum level for the purposes of this requirement.[88]
3. The Secretary must certify that the amount available to the agency for anticipated and ongoing suppression operations for the wildfire on which the declaration request has been made will be obligated not later than 30 days after the notification.
4. The Secretary must specify the amount of current fiscal year funds that are required for suppression operations related to the wildfire for which the declaration is requested.

Title IX of H.R. 2647 would require that the President establish a specific account from which available funds may be transferred to the DOI Secretary or USDA Secretary to conduct wildfire suppression operations if a declaration of a major disaster is made.[89] This account may be used only to fund assistance for major disaster declarations for wildfires on federal lands and nonfederal lands pursuant to a fire protection agreement or cooperative agreement. The bill would establish reimbursement procedures from the nonfederal entity if any of the funds transferred from this account were used for wildfires on nonfederal lands under a fire protection agreement.

This account could be funded in anticipation of one or more such major disasters, or in response to them, through regular or supplemental appropriations.

Within 90 days after the end of a fiscal year in which a major disaster for wildfire on federal lands has been declared, the affected Secretaries must submit reports detailing their wildland fire management decisions and cost expenditures to several congressional committees. These reports must include the risk-based factors that influenced management decisions; an analysis of the effectiveness of management decisions and cost factors across a statistically significant sample of large fires; suppression operations expenditures broken out by fire size, cost, regional location, and other factors as deemed appropriate by each Secretary; and lessons learned.

Section 903 of H.R. 2647 would prohibit the agencies from transferring funds from their other accounts for suppression purposes, potentially eliminating the agencies' fire borrowing.

Potential Implications

The budgetary mechanism under this proposal potentially could provide an additional discretionary spending option for wildfire suppression operations funding related to the disaster relief designation. As previously discussed, spending designated as "disaster relief," which is effectively exempt from the BCA limits, may be provided only for activities carried out pursuant to a determination of a major disaster under Section 102(2) of the Stafford Act.[90] Because H.R. 2647 would amend the Stafford Act to include wildfires on federal land as an incident for which major disasters could be declared, these changes might allow appropriations for such activities to be disaster-designated under certain circumstances:

- First, the preconditions for requesting a major disaster declaration for wildfires on federal land would need to be fulfilled. These preconditions include a minimum level of budgetary resources for wildfire suppression operations, and that minimum amount would need to have been appropriated (and therefore would be subject to the statutory discretionary spending limits).
- Second, even if disaster-designated funds were appropriated in anticipation of a major disaster for wildfires on federal lands, they could be transferred and made available to the requesting agency only

> after a major disaster had been declared by the President. They would not be available to the agencies prior to that time.

In the contexts of their various budget enforcement responsibilities, however, the House and Senate Budget Committees, the Congressional Budget Office, and the Office of Management and Budget would need to determine how the changes to the Stafford Act proposed by H.R. 2647 would affect the BBEDCA disaster designation, if at all.

If the Stafford Act changes in Title IX of H.R. 2647 led to the use of the disaster designation for additional wildfire suppression operations funding, these changes would have implications for the future operation of that designation. In addition to allowing wildfires that meet certain criteria to be eligible for funding using the disaster designation, the calculation for the disaster cap would only gradually incorporate appropriations for any major disaster declarations for wildfires on federal lands over the next 10 fiscal years. This has the potential to increase the competition for budgetary resources that can be provided using the designation. However, because appropriations for the full adjustment allowed by the disaster cap have not been made since FY2013, adding purposes to the designation might cause more of the allowable adjustment to be used and factored into future disaster cap calculations. This might have the effect of increasing the amount of disaster cap in future fiscal years above what it otherwise might have been. Because the amount available under the disaster designation is highly sensitive to future appropriations decisions, CRS cannot determine definitively the extent to which the calculation for the disaster cap would be affected by this expansion of Stafford Act purposes.

Other considerations also affect the budgetary implications of this proposal for DOI and FS. The bill would prohibit fire borrowing, although it is unclear if fire borrowing authorities would continue to be provided in appropriations laws. It is also unclear how H.R. 2647 would impact the agencies' budget formulation process. The precondition to appropriate a minimum amount is based on the rolling 10-year suppression obligation average, but the rescissions and unobligated balances might complicate that calculation.

H.R. 2647 would require the agencies to provide Congress and the public with reports analyzing cost expenditures and management decisions. These reporting requirements could provide Congress—and the agencies—with more information on agency decisionmaking and spending to aid in future appropriations decisions and oversight. However, the development and production of these reports and the required analyses potentially could require substantial investments from the agencies.

Legislative Action

Title IX was not in the versions of H.R. 2647 that were reported from the House Committees on Agriculture and Natural Resources.

On July 7, 2015, the House Committee on Rules reported a special rule (H.Res. 347) for the consideration of H.R. 2647. The special rule provides that Rules Committee Print 114-24, as modified by an amendment printed in the Rules Committee report on H.Res. 347, shall be considered to be the base text of the bill for the purposes of amending it. (The modifications to the committee print do not alter Title IX.)

End Notes

[1] National Wildfire Coordinating Group, "Glossary of Wildland Fire Terminology," October 2014, at http://www.nwcg.gov/pms/pubs/glossary/w.htm. For a more comprehensive discussion on wildland fire management, see CRS Report RL30755, *Forest Fire/Wildfire Protection*, by Kelsi Bracmort.

[2] Other federal agencies—such as the Department of Defense—are responsible for wildfire response on their lands. This report focuses on Forest Service (FS) and Department of the Interior (DOI) wildfire management responsibilities.

[3] National Interagency Fire Center (NIFC), Fire Executive Council, *Guidance for Implementation of Federal Wildland Fire Policy*, February 13, 2009, p. 10, at http:// www.nifc.gov/ policies/ policies_main.html.

[4] For more information, see "Federal Assistance for Nonfederal Wildfires" in this report, and CRS Report R43738, *Fire Management Assistance Grants: Frequently Asked Questions*, coordinated by Bruce R. Lindsay.

[5] NIFC, "Wildland Fire Statistics," at http://www.nifc.gov/fireInfo/fireInfo_statistics.html.

[6] NIFC National Interagency Coordination Center, *Wildland Fire Summary and Statistics, 2014*, at http://www.predictiveservices.nifc.gov/intelligence/2014_Statssumm/ 2014Stats&Summ.html.

[7] The West includes Alaska, Arizona, California, Colorado, Hawaii, Idaho, Montana, New Mexico, Nevada, Oregon, Utah, Washington, and Wyoming.

[8] V. C. Radeloff et al., "The Wildland-Urban Interface in the United States," *Ecological Applications*, vol. 15, no. 3 (2005), pp. 799-805. For more information on the WUI, see CRS Report RS21880, *Wildfire Protection in the Wildland- Urban Interface*, by Katie Hoover and Kelsi Bracmort.

[9] Forest Service, *Wildfire, Wildlands, and People: Understanding and Preparing for Wildfire in the Wildland-Urban Interface*, GTR-299, January 2013.

[10] CoreLogic, Inc, *Wildfire Hazard Risk Report*, 2015.

[11] NIFC, *Guidance*, p. 11.

[12] NIFC, National Multi-Agency Coordinating Group, *NMAC National Strategy*, 2013, at http:// www.nifc.gov/nicc/ administrative/nmac/index.html.

[13] See for example, National Academy of Public Administration (NAPA), *Wildfire Suppression: Strategies for Containing Costs*, September 2002; Dean Lueck, "Economics and the

Organization of Wildfire Suppression," in *Wildfire Policy: Law and Economics Perspectives*, ed. Karen M. Bradshaw and Dean Lueck (New York: RFF Press, 2012); Randal O'Toole, *Reforming the Fire Service: An Analysis of Federal Fire Budgets and Incentives*, Thoreau Institute, July 2002; and Timothy Ingalsbee, *Getting Burned: A Taxpayer's Guide to Wildfire Suppression Costs*, Firefighters United for Safety, Ethics, & Ecology, August 2010.

[14] J. Liang et al., "Factors influencing Large Wildland Fire Suppression Expenditures," *International Journal of Wildland Fire*, vol. 17 (2008), pp. 650-659.

[15] USDA Office of Inspector General (OIG), *Audit Report: Forest Service Large Fire Suppression Costs*, Report No. 08601-44-SF, 2006.

[16] P. H. Gude et al., "Evidence for the Effect of Homes on Wildfire Suppression Costs," *International Journal of Wildland Fire*, vol. 22, no. 4 (2013), pp. 537-548; K. M. Gebert, D.E. Calkin, and J. Yoder, "Estimating Suppression Expenditures for Individual Large Wildland Fires," *Western Journal of Applied Forestry*, vol. 22, no. 3 (2007), pp. 188-196.

[17] U.S. Government Accountability Office (GAO), *Wildland Fire Suppression: Lack of Clear Guidance Raises Concerns about Cost Sharing between Federal and Nonfederal Entities*, GAO-06-570, May 2006; USDA OIG, *Audit Report*; NAPA, *Wildfire Suppression*; O'Toole, *Reforming the Fire Service*; Ingalsbee, *Getting Burned*; and Headwaters Economics, *Solutions to the Rising Costs of Firefighting in the Wildland-Urban Interface*, September 2009.

[18] GAO, *Wildland Fire Management: Lack of Clear Goals or a Strategy Hinders Federal Agencies' Efforts to Contain Costs*, GAO-07-655, June 1, 2007.

[19] GAO-06-570.

[20] For more information, see the National Interagency Coordination Center website at http://www.nifc.gov/nicc/ and the Geographic Area Coordination Centers website at http://gacc.nifc.gov/.

[21] 42 U.S.C. §5122(2). For more information on declarations, see CRS Report RL33053, *Federal Stafford Act Disaster Assistance: Presidential Declarations, Eligible Activities, and Funding*, by Francis X. McCarthy; CRS Report R43784, *FEMA's Disaster Declaration Process: A Primer*, by Francis X. McCarthy; and CRS Report R42702, *Stafford Act Declarations 1953-2011: Trends and Analyses, and Implications for Congress*, by Bruce R. Lindsay and Francis X. McCarthy.

[22] See http://www.fema.gov/disasters.

[23] 42 U.S.C. §5191.

[24] 42 U.S.C. §5187.

[25] For more information, see CRS Report R43738, *Fire Management Assistance Grants: Frequently Asked Questions*, coordinated by Bruce R. Lindsay.

[26] FEMA was unable to provide data on Fire Management Assistance Grant (FMAG) request denials.

[27] S. 2534, §573, pp. 106-107.

[28] P.L. 114-4, §570. The provision in the appropriations bill was similar to legislation introduced during the 113th and 114th Congresses. These bills included H.R. 3333, the Wildfire Prevention Act of 2013 (S. 1396), which authorized mitigation for FMAGs, and the PREPARE Act (S. 1428), which did the same.

[29] See CRS Report R43537, *FEMA's Disaster Relief Fund: Overview and Selected Issues*, by Bruce R. Lindsay.

[30] Wildfire appropriations to DOI used to go directly to BLM and were then allocated among the other bureaus, but since 2009 appropriations have gone to the DOI department-level Office of Wildland Fire for allocation.

[31] For background information on the wildland fire accounts for DOI and FS, see CRS Report R43077, *Wildfire Management: Federal Funding and Related Statistics*, by Katie Hoover and Kelsi Bracmort.

[32] P.L. 111-88, Division A, Title V (43 U.S.C. §1748a).

[33] Appropriations to the Preparedness and Suppression programs are fungible, meaning the funds are interchangeable. For example, maintenance and storage for aviation assets may be funded through the preparedness activity until activated for wildfire response, at which point suppression funds will cover operation expenses.

[34] Title V of Division A of the FY2010 Department of the Interior, Environment, and Related Agencies Appropriations Act, P.L. 111-88 (43 U.S.C. §§1748a et seq.).

[35] H.Rept. 111-316.

[36] 16 U.S.C. §1748a(e).

[37] DOI, *FY2016 Wildland Fire Management Budget Justification*.

[38] Email from FS Legislative Affairs staff, May 2015.

[39] In the years that fire transfers did not occur but the FS WFM suppression and FLAME accounts were exhausted, FS used unobligated balances from previous fiscal years to cover additional suppression expenses as needed. Email from FS Legislative Affairs staff, May 2015.

[40] Email from DOI, May 2015.

[41] The provision generally reads: "Any appropriations or funds available to the Forest Service may be transferred ... for forest firefighting and the emergency rehabilitation of burned-over lands under its jurisdiction."

[42] In general, the agencies will have already depleted their WFM suppression accounts and transferred funds from their FLAME reserve accounts.

[43] Email from FS Legislative Affairs staff, February 2015.

[44] Historically, the FS borrowed funds primarily from its mandatory spending accounts, particularly the Knutson- Vandenberg (K-V) Fund. This account accumulated deposits from timber purchasers to reforest and otherwise improve timber in timber sale areas. Because of the lag between timber payments and reforestation, the K-V Fund often had a balance of about $500 million—more than enough to borrow for emergency fire suppression without impinging on one season's tree planting efforts. However, the K-V Fund has had a smaller balance since FY2000 (because of lower timber sales) while emergency wildfire suppression costs have risen. Thus, the FS has had to borrow funds from other FS accounts—land and easement purchases, recreation and wildlife management, and more.

[45] GAO, *Wildland Fire Management: Actions by Federal Agencies Could Mitigate Rising Fire Costs and Their Effects on Other Agency Programs*, GAO-09-444T, June 2004.

[46] 43 U.S.C. §1748a(2)(C)(ii) states that FS and DOI "should promptly make a supplemental request for additional funds to replenish the FLAME Fund if the Secretary determines that the FLAME Fund will be exhausted within 30 days."

[47] See, for example, the following GAO reports on wildland fire funding issues published between 2004 and 2009: GAO, *Wildfire Suppression: Funding Transfers Cause Project Cancellations and Delays, Strained Relationships, and Management Disruptions*, GAO-04-612, June 2004; GAO-07-655; and GAO-09-444T.

[48] National Academy of Public Administration, *Wildfire Suppression: Strategies for Containing Costs*, September 2002.

[49] See for example, GAO-04-612; testimony of FS Chief Tom Tidwell, in U.S. Congress, Senate Committee on Energy and Natural Resources, *Hearing to Receive Testimony on the Federal Government's Role in Wildfire Management, the Impact of Fires on Communities, and Potential Improvements to Be Made in Fire Operations*, 114th Cong., 1st sess., May 5, 2015; and FS, *Fire Transfer Impact by State*, June 9, 2014, at http:// www.fs.fed.us/ publications/ forest-service- fire-transfer-state-impacts.pdf.

[50] GAO-04-612.

[51] H.Rept. 111-316.

[52] FS, *FY2016 Budget Justification*, p. 400, at http://www.fs.fed.us/about-agency/budget-performance.

[53] For example, "designated by the Congress as being for an emergency requirement pursuant to §251(b)(2)(A)(i) of the Balanced Budget and Emergency Deficit Control Act of 1985." For

more information on emergency designations, see CRS Report R41564, *Emergency Designation: Current Budget Rules and Procedures*, by Bill Heniff Jr.

[54] See FS and DOI annual budget justification documents for a description of how the rolling 10-year suppression obligation is calculated.

[55] U.S. Congress, House Committee on Natural Resources, *Federal Land Assistance, Management and Enhancement Act*, Report to accompany H.R. 5541, 110th Cong., 2nd sess., June 10, 2008, H.Rept. 110-704.

[56] For more information on the Administration's proposal, see FS FY2016 Budget Justification, p. 251; and DOI Wildland Fire Management FY2016 Budget Justification, p. 35.

[57] In 2004, GAO recommended that FS and DOI develop a method to predict suppression spending that was more accurate than using the rolling 10-year obligation average (GAO-04-612).

[58] GAO-04-612.

[59] GAO-09-444T, p. 8.

[60] See §2(c)(2) of H.R. 5541 from the 110th Congress.

[61] 43 U.S.C. §1748a(h)(3)(D).

[62] 43 U.S.C. §1748a(h)(3)(D).The FS Southern Research Station runs the regression models for both FS and DOI based on methods developed by J. P. Prestemon, K. L. Abt, and K. Gebert, "Suppression Cost Forecasts in Advance of Wildfire Seasons," *Forest Science*, vol. 54, no. 4 (2008), pp. 381-396; and K. L. Abt, J. P. Prestemon, and K. Gebert, "Wildfire Suppression Cost Forecasts for the U.S. Forest Service," *Journal of Forestry*, vol. 107, no. 4 (2009), pp. 173- 178.

[63] A prior version of the FLAME Act was first introduced in the 109th Congress. The 109th and 110th Congresses also considered other bills to address wildfire spending, such as S. 3729 [109th Congress] and S. 1770 [110th Congress]. The Administration proposed the establishment of a suppression reserve fund in FY2010 and FY2011, which was similar to the FLAME Act but with some fundamental differences, including requiring a presidential declaration to access the funds. See FS's FY2011 Budget Justification, p. 13-1, for more information on the proposal.

[64] H.Rept. 111-316.

[65] GAO-09-444T; D. E. Calkin, M. P. Thompson, and M. A. Finney, "Negative Consequences of Positive Feedbacks in U.S. Wildfire Management," *Forest Ecosystems*, vol. 2, no. 9 (2015).

[66] See for example, Calkin et al., "Negative Consequences"; GAO, *Wildland Fire Management: Lack of a Cohesive Strategy Hinders Agencies' Cost-Containment Efforts*, GAO-07-427T, January 2007; GAO, *Wildland Fire Management: Federal Agencies Have Taken Important Steps Forward, but Additional, Strategic Action Is Needed to Capitalize on Those Steps*, September 2009.

[67] Ingalsbee, *Getting Burned*; and Headwaters Economics, *Solutions to the Rising Costs of Firefighting in the Wildland- Urban Interface*, September 2009.

[68] P.L. 112-25.

[69] For further information with regard to budget functions, see CRS Report 98-280, *Functional Categories of the Federal Budget*, by Bill Heniff Jr.

[70] The sequestration mechanism to enforce the statutory discretionary spending limits is provided by the BBEDCA, as amended by the BCA. For further information about these procedures, see CRS Report R41965, *The Budget Control Act of 2011*, by Bill Heniff Jr., Elizabeth Rybicki, and Shannon M. Mahan. The Congressional Budget Act of 1974 (Titles I-IX of P.L. 93-344, 2 U.S.C. §601-688) provides for procedural enforcement consistent with the statutory spending limits, including the adjustments to those limits.

[71] Title II of P.L. 99-177, 2 U.S.C. §900-922, as amended by the BCA.

[72] Although Congress may apply the emergency designation at its discretion, the BBEDCA, §250(c)(20), defines *emergency* to mean "a situation that—(A) requires new budget authority and outlays (or new budget authority and the outlays flowing therefrom) for the

prevention or mitigation of, or response to, loss of life or property, or a threat to national security; and (B) is unanticipated."

[73] 42 U.S.C. §5122(2).

[74] For additional discussion on the BCA and disaster spending, see CRS Report R42352, *An Examination of Federal Disaster Relief Under the Budget Control Act*, by Bruce R. Lindsay, William L. Painter, and Francis X. McCarthy.

[75] BBEDCA, §251(b)(2)(D). The amount that is allowed for the disaster designation each fiscal year is calculated by the Office of Management and Budget.

[76] The DRF provides funding for all open disasters, disasters currently impacting the nation, and retains a balance for possible future disaster events.

[77] U.S. Department of Homeland Security, *Federal Emergency Management Agency Disaster Relief Fund: FY2013 Congressional Justification*, p. 5, at http:// www.fema.gov/ pdf/ about/ budget/ 11f_fema_disaster_relief_fund_dhs_fy13_cj.pdf.

[78] For further information about S. 1875 and H.R. 3992 (113th Cong.), see the CRS congressional distribution memorandum "The Wildfire Disaster Funding Act (S. 1875 and H.R. 3992), 113th Congress, March 26, 2014, available from the authors.

[79] The BCA spending limits are currently through FY2021.

[80] 43 U.S.C. §1748(a) (g).

[81] For more information, see FS FY2016 Budget Justification, p. 251; and DOI Wildland Fire Management FY2016 Budget Justification, p. 35.

[82] S.Con.Res. 11, §319. For information about reserve funds, see CRS Report R43535, *Provisions in the Bipartisan Budget Act of 2013 as an Alternative to a Traditional Budget Resolution*, by Megan S. Lynch.

[83] S.Con.Res. 11, §424.

[84] This includes activities authorized by the Healthy Forests Restoration Act of 2003 (16 U.S.C. §§6501 et seq.) and the Tribal Forest Protection Act of 2004 (25 U.S.C. §3115a).

[85] Fire regime condition class is a classification that describes the relative change between the historical (prior to modern human intervention) frequency and intensity of fire patterns across a vegetated landscape to the current fire patterns. For more information, see S. Barrett et al., *Interagency Fire Regime Condition (FRCC) Guidebook Version 3.0*, 2010, at https://www.frames.gov/frcc.

[86] See U.S. Congress, House Committee on Rules, *Text of H.R. 2647, Resilient Federal Forests Act of 2015*, committee print, 114th Cong., 1st sess., July 1, 2015, H.Prt. 114-21.

[87] Currently, "fire" is an eligible event under the Stafford Act. This proposal extends that authority to federal lands.

[88] It is not clear as of what date during the fiscal year that the amount of these unobligated balances are to be calculated for the purposes of this precondition.

[89] In the case of each declaration for which a transfer of funds is to occur, the President has the authority to decide where to establish the account. In addition, the amount of the transfer is limited to the amount that was requested by the agency as part of the request for the disaster declaration, and shall not exceed the amount in the account.

[90] Section 251(b) (2)(D) of the BBEDCA.

In: Wildfire Suppression
Editor: Neal Jenkins
ISBN: 978-1-63483-773-6

Chapter 2

THE RISING COST OF WILDFIRE OPERATIONS: EFFECTS ON THE FOREST SERVICE'S NON-FIRE WORK*

United States Forest Service

OVERVIEW

Over 100 years ago, President Theodore Roosevelt established the U.S. Forest Service to manage America's 193-million acre national forests and grasslands for the benefit of all Americans. Today, that mission is being consumed by the ever-increasing costs of fighting fires.

This report documents the growth over the past 20 years of the portion of the Forest Service's budget that is dedicated to fire, and the debilitating impact those rising costs are having on the recreation, restoration, planning, and other activities of the Forest Service.

In 1995, fire made up 16 percent of the Forest Service's annual appropriated budget—this year, for the first time, more than 50 percent of the Forest Service's annual budget will be dedicated to wildfire.[1] Along with this shift in resources, there has also been a corresponding shift in staff, with a 39 percent reduction in all non-fire personnel. Left unchecked, the share of the budget devoted to fire in 2025 could exceed 67 percent, equating to reductions of nearly $700 million from non-fire programs compared to today's funding

---

* This is an edited, reformatted and augmented version of a report issued August 4, 2015.

levels. That means that in just 10 years, two out of every three dollars the Forest Service gets from Congress as part of its appropriated budget will be spent on fire programs.

As more and more of the agency's resources are spent each year to provide the firefighters, aircraft, and other assets necessary to protect lives, property, and natural resources from catastrophic wildfires, fewer and fewer funds and resources are available to support other agency work—including the very programs and restoration projects that reduce the fire threat.

The depletion of non-fire programs to pay for the ever-increasing costs of fire has real implications, not only for the Forest Service's restoration work that would help prevent catastrophic fires, but also for the protection of watersheds and cultural resources, upkeep of programs and infrastructure that support thousands of recreation jobs and billions of dollars of economic growth in rural communities, and support for the range of multiple uses, benefits and ecosystem services, as well as research, technical assistance, and other programs that deliver value to the American public.

The Forest Service has continually worked to do more with less, seeking to provide for the forests' multiple uses with fewer resources and staff. The Forest Service has also worked to appropriately allocate firefighting resources and improve risk management to use those resources safely and efficiently.

However, the agency is at a tipping point.

Climate change has led to fire seasons that are now on average 78 days longer than in 1970. The U.S. burns twice as many acres as three decades ago and Forest Service scientists believe the acreage burned may double again by mid-century. Increasing development in fire-prone areas also puts more stress on the Forest Service's suppression efforts.

While the Forest Service and its firefighting partners are able to suppress or manage 98 percent of fires, catastrophic mega-fires burn through the agencies resources: 1–2 percent of fires consume 30 percent or more of annual costs. Last year, the Forest Service's 10 largest fires cost more than $320 million dollars. The cost of fire suppression is predicted to increase to nearly $1.8 billion by 2025. This trend of rising fire suppression costs is predicted to continue as long as the 10-year average serves as the funding model and presents a significant threat to the viability of all other services that support our national forests.

This unsustainable problem is made worse because in many years, fighting fires costs more than was planned for that year, requiring mid-season transfers of additional dollars from already depleted accounts to pay for firefighting: a practice referred to as "fire transfer." In some cases, the agency is forced to divert money away from the same forest restoration projects that prevent or lessen the impacts of future wildfire. While Congress typically provides supplemental resources to replenish the Forest Service budget after fire transfers, transfers remain extremely problematic as they disrupt seasonal work, frustrate partners, and delay vital work.

THE ESCALATING COST OF FIRE SUPPRESSION

Wildland fire suppression activities are currently funded entirely within the U.S. Forest Service budget, based on a 10-year rolling average. Using this model, the agency must average firefighting costs from the past 10 years to predict and request costs for the next year. When the average was stable, the agency was able to use this model to budget consistently for the annual costs associated with wildland fire suppression. Over the last few decades, however, wildland fire suppression costs have increased as fire seasons have grown longer and the frequency, size, and severity of wildland fires has increased.

Changing climatic conditions across regions of the United States are driving increased temperatures - particularly in regions where fire has not been historically prominent. This change is causing variations and unpredictability in precipitation and is amplifying the effects and costs of wildfire. Related impacts are likely to continue to emerge in several key areas: limited water availability for fire suppression, accumulation at unprecedented levels of vegetative fuels that enable and sustain fires, changes in vegetation community composition that make them more fire prone, and an extension of the fire season to as many as 300 days in many parts of the country.

These factors result in fires that increasingly exhibit extreme behavior and are more costly to manage. The six worst fire seasons since 1960 have all occurred since 2000. Moreover, since 2000, many western states have experienced the largest wildfires in their state's history.

In addition, more and more development is taking place near forests—an area referred to as the Wildland-Urban Interface (WUI). Increasing densities of people and infrastructure in the WUI makes management more complex and requires more firefighting assets to ensure an appropriate, safe, and effective response that protects lives and property.

Funding for non-fire programs has not kept pace with the increased cost of fighting fire. The growth in fire suppression costs has steadily consumed an ever-increasing portion of the agency's appropriated budget. Between last fiscal year and this year, for example, the suppression budget grew by $115 million and non-fire programs were reduced by that amount, requiring the agency to forego opportunities to complete vital restoration work and meet public expectations for services. Those non-fire activities are often those that improve the health and resilience of our forested landscapes and mitigate the potential for wildland fire in future years.

Over the last few decades, wildfire costs have increased as a percent of the Forest Service's budget as fire seasons have grown longer and more costly. The projected continued growth in the 10-year average cost of fire suppression through 2025 is rising to nearly $1.8 billion. This amounts to a nearly $700 million decrease in non-fire program funding in the next 10 years.

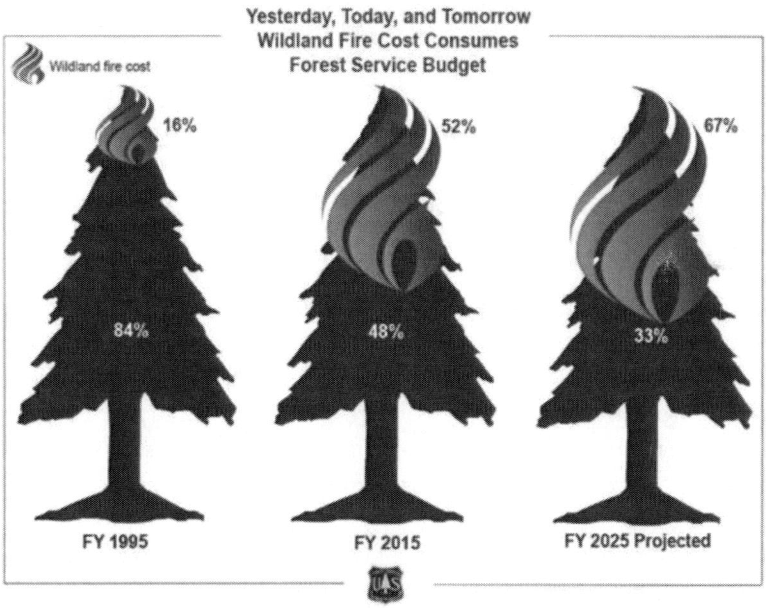

Figure 1. The Cost of Wildland Fire (Preparedness, Suppression, FLAME, and related programs) as a Percentage of the Forest Service's Annual Budget.

Impacts to Forest Service Program Areas and Staff

The following charts show that from 1995 to 2015, the Wildland Fire Management appropriation (Preparedness, Suppression, FLAME, and related programs) has more than tripled in its portion of the Forest Service budget from 16 percent to 52 percent, reducing National Forest System funding by nearly $475 million in 2015 dollars (32 percent reduction in real dollars), and also impacting other program areas.

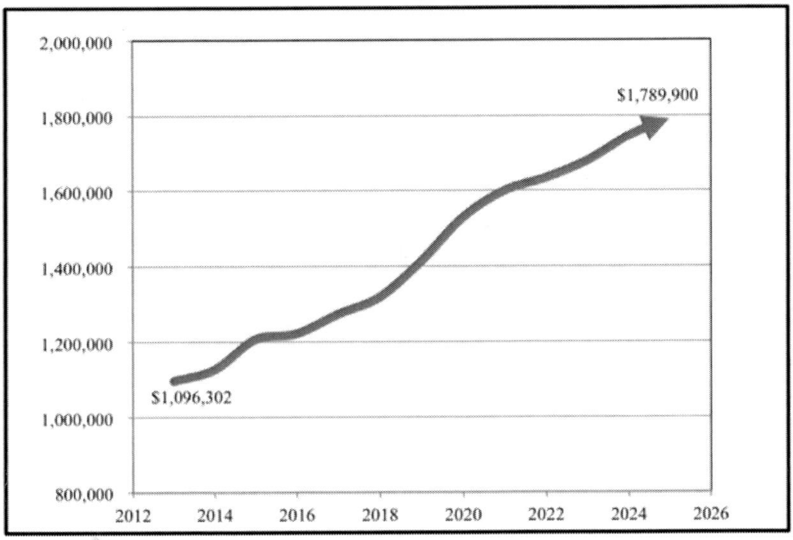

Figure 2. Projected Growth of the 10-Year Average Cost of Fire Suppression (in $1000s) Through 2025.[2]

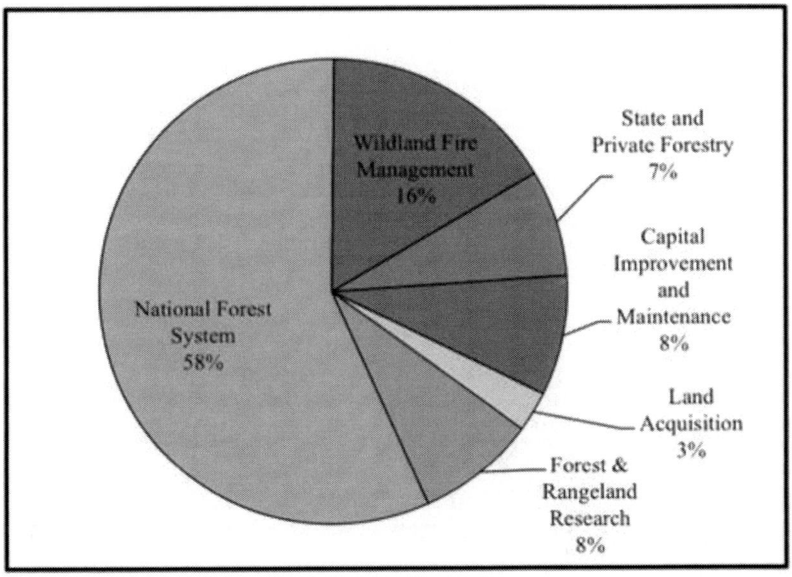

Figure 3. Forest Service FY 1995 Appropriations by Fund.

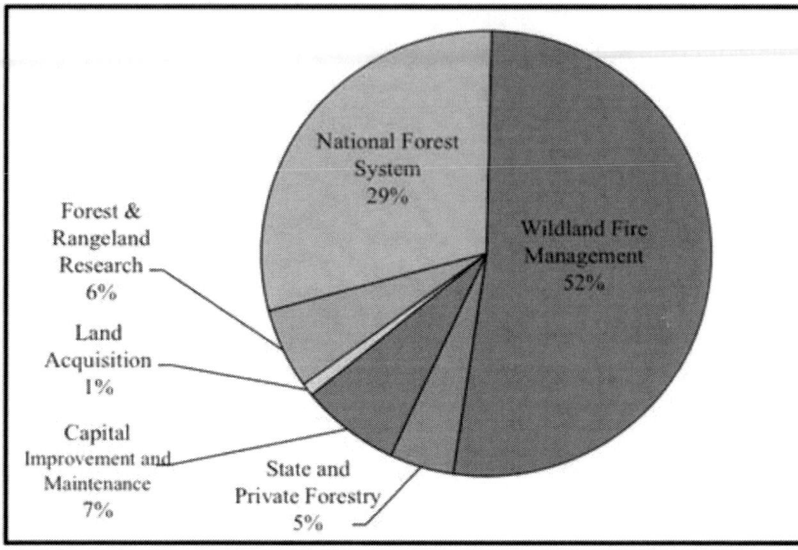

Figure 4. Forest Service FY 2015 Appropriations by Fund.

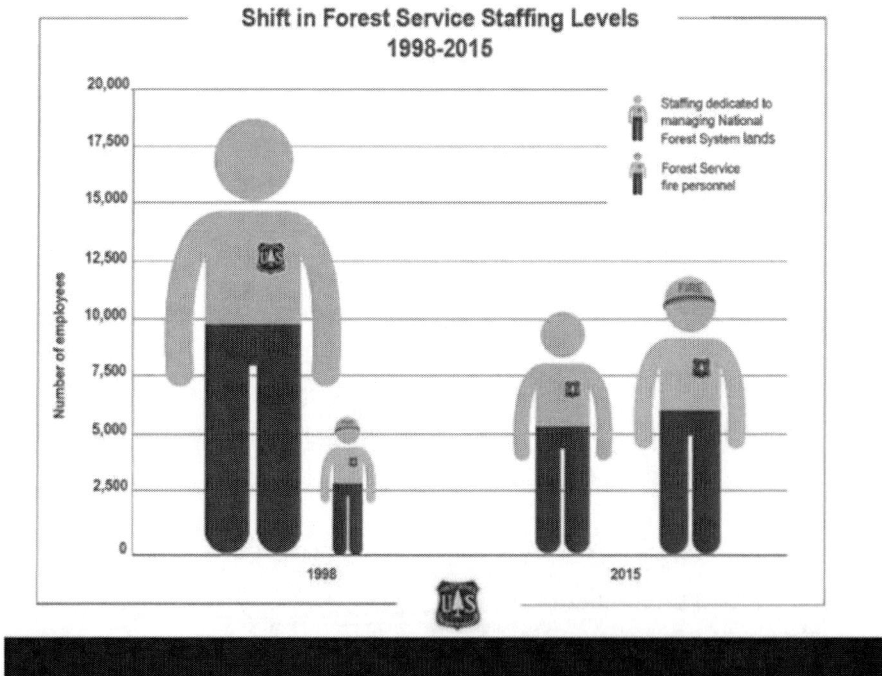

Figure 5. Forest Service staffing 1998 to 2015.

At the same time, increasing the portion of the budget dedicated to fire has reduced the Forest Service's ability to sustain staffing in vital non-fire program areas, which negatively impacts the Forest Service's ability to deliver work on the ground, including forest restoration and management, recreation, research, watershed protection, land conservation, and other activities.

Since 1998, fire staffing within the Forest Service has increased 114 percent, from around 5,700 employees in 1998 to over 12,000 in 2015. Over the same period, staffing levels for those dedicated to managing National Forest System lands has decreased by 39 percent—from approximately 18,000 in 1998 to fewer than 11,000 in 2015.

IMPACTS ON INDIVIDUAL PROGRAMS

Funding trends for the past 15 years for individual programs used to manage the National Forest System are shown below.[3] This section illustrates the significant declines in these programs and describes the resulting impacts as a result of the shift of financial and human resources away from management of the National Forest Systems and to the Wildland Fire Management accounts, within the agency's constrained annual budget.[4]

Vegetation & Watershed Management—24% Reduction

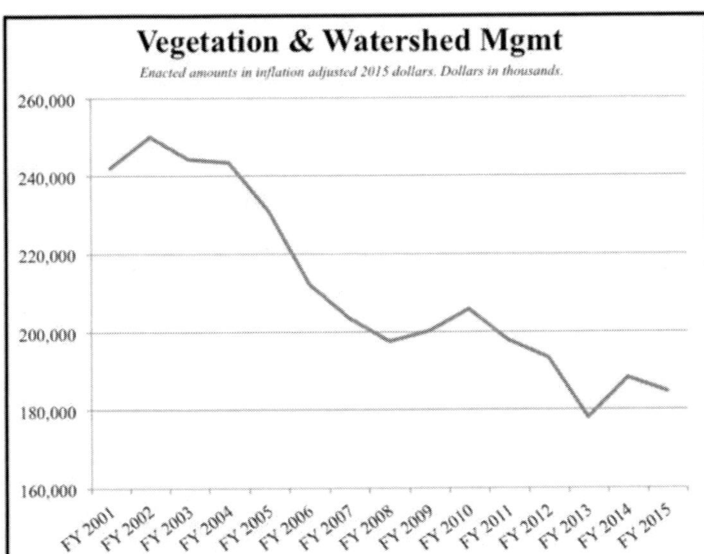

The Vegetation and Watershed Management Program is the cornerstone for forest, rangeland, soil and water restoration and enhancement activities on National Forest System (NFS) lands and plays a key role in post-fire restoration. These programs are necessary for the agency to effectively support resource restoration projects that achieve multiple values, develop external partnerships to sustain healthy watersheds and ecological communities, and provide an array of benefits for current and future generations.

Benefits include: improved water quality and quantity, healthy forests and rangelands that provide a variety of products, a reduction of risks associated with wildfires, and greater resistance to establishment and spread of invasive species.

Reduced funding since 2001 has decreased the rate of restoration that the agency could have achieved across all NFS landscapes had funding levels been maintained. The agency has been less able to engage in watershed improvement activities as well as overall restoration activities occurring within priority watersheds. The reductions have limited the agency's ability to prevent and limit the spread of invasive species; to decrease the backlog of deforestation and young stand management needs; and to lessen the severity and extent of insect, disease, and fire-prone forest stands.

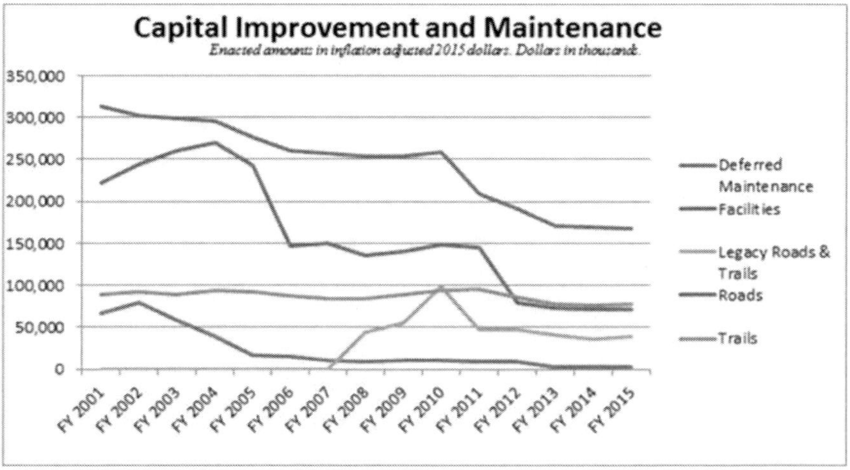

Facilities—68% Reduction

The facilities program supports maintenance and capital improvement on approximately 21,600 recreation sites and 23,100 research and other administrative buildings. More than half of all administrative facilities need improvement, with approximately 41 percent in poor condition needing major repairs or renovation, approximately 12 percent in fair condition needing some minor repair, and 47 percent of the facilities in good condition.

Due to the significant decrease in facilities funding, the agency has had to scale down or defer most decommissioning and disposal projects that would reduce our square footage. Projects that implement sustainability best practices to conserve energy and water have been deferred. Reduced funding has jeopardized the agency's ability to address basic facility operational and maintenance needs and many of our safety issues such as those associated with water and septic systems. The overall effect is an increase in public health and safety concerns, and liability for the federal government. The only action national forests can take to reduce the government liability is to close recreation facilities, thereby impacting the outdoor recreation opportunities that drive many rural tourism economies.

Because of a lack of funding, the Forest Service has lost opportunities for new office construction to replace administrative facilities at the end of their design life, resulting in office closure and moves into leased facilities. Deferred maintenance has increased resulting in more expensive future repairs or possible loss of facility investments. Projects to decommission buildings and reduce the square feet have been deferred. Projects that would have helped the agency make its buildings more sustainable and reduce the costs of maintaining recreation facilities have also been deferred.

Concurrently, additional fire facilities have been added to meet increased suppression needs. This has required a shift in spending from capital improvements for national priority projects to projects critical for fire readiness. For example, Air Tanker Bases need pavement improvements to handle the newer aircraft which are being added to the fleet, these projects will be phased in over several years, increasing costs.

Roads—46% Reduction

The National Forest Road System is an integral part of the rural transportation network. It provides access for recreational, administrative, resource management, and commercial purposes. It also provides access to and between rural and gateway communities, contributing to community vitality and economic development. Maintaining this system is necessary to continue to provide this access, to meet Highway Safety Act requirements, emergency response, and to protect the quality of critical water supplies provided by National Forest System lands to communities.

As our transportation infrastructure ages or is damaged by natural events, some roads and bridges have become unsafe for public travel. The Forest Service has had to restrict traffic on or close those roads and bridges until funds are available for maintenance and repairs. Thirteen percent of our bridges are currently structurally deficient and the average age of all bridges is 50 years old. Without needed replacement or repairs, structurally deficient bridges would first be load-restricted and ultimately closed as deficiencies progress.

Road restoration and decommissioning efforts, a critical component of watershed restoration, has progressed but not at the accelerated pace necessary to achieve watershed protection and other management objectives. Action to replace road and stream crossings for aquatic organism passage and to improve aquatic habitat and resilience to catastrophic natural disasters has also suffered. Reducing these activities increases the severity and frequency of environmental impacts like catastrophic failure from natural disasters, adverse effects to water quality, and deterioration of aquatic and terrestrial habitat. Our ability to support accelerated watershed restoration continues to be severely affected.

As a lack of funds forces the Forest Service to delay needed maintenance and improvements on many roads and bridges, access will become more restricted, environmental impacts will increase rapidly, and vulnerability to catastrophic failure from natural disasters will greatly increase. It will become more difficult for the public to access communities reached only by traveling through NFS lands, as well as recreational areas, and other natural resource areas. Firefighting ability could also decline significantly as road access restrictions increase, putting our resources and the public at risk.

Deferred Maintenance—95% Reduction

The Deferred Maintenance and Infrastructure Improvement program addresses serious public health and safety concerns associated with the agency's backlog in maintenance needs. This program funds high-priority national projects, focusing on areas that are heavily used by the public and agency employees. These projects include critical maintenance and repairs to dams; correcting health and safety deficiencies in buildings, campgrounds, and water and wastewater systems; and renovating recreation structures.

In FY 2001, the Deferred Maintenance funding supported approximately 400 major projects. In FY 2014 the funding supported three major projects. In FY 2013 and FY 2014, 21 projects were deferred to future years, including sewer system repairs, water system improvements, dam repairs, and wastewater system rehabilitation.

Forest Service assets currently have a deferred maintenance backlog of over $5.1 billion and many are 30 to 50 years old or more. The near elimination of funding for this program has prevented the Forest Service from making a dent in this backlog and has forced the agency to shift more of the deferred maintenance work to other capital improvement programs, further reducing our ability to improve the long-term sustainability of our facilities, roads, and trails.

Recreation, Heritage and Wilderness—15% Reduction

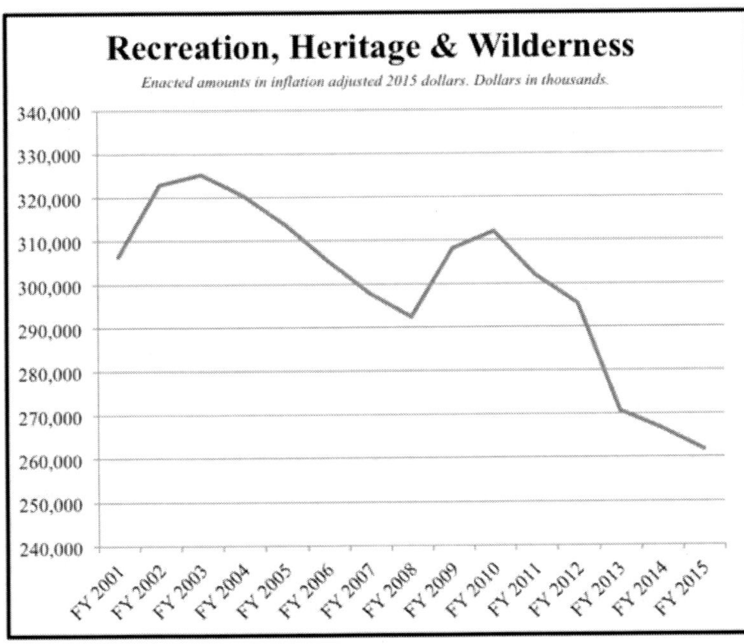

The Recreation, Heritage and Wilderness programs offer a diverse range of recreational opportunities across NFS lands, connecting people with nature in an unmatched variety of settings and activities. The decrease in funding resulting from increased fire costs has limited the agency's ability to provide vital recreational opportunities on NFS lands, which jeopardizes the thousands of jobs that are part of a growing recreational economy.

The agency has been unable to more fully implement sustainable Recreation, Heritage, Volunteer Services and Wilderness and Wild & Scenic Rivers programs to provide consistent, quality recreation opportunities to the public. Reductions in recreation funding have a direct impact on local economies supported by these activities, including many small outfitter and guide businesses that depend on recreation sites and programs on NFS lands. Additionally, the Forest Service's ability to leverage funds and implement projects with partners and volunteers is constrained by the reductions in funding and staff, substantially affecting services.

The reductions in funding have also affected the Forest Service's capacity to manage the permits needed for outfitters and guides and other recreation-focused small businesses to use the public land. This impacts the presence and stability of permittees and small business in nearby tourism-oriented communities.

Services to youth have also suffered. A higher sustained level of funding would increase the capability of the Forest Service to engage youth in the outdoors and support increased employment opportunities for youth and veterans through programs such as the 21st Century Conservation Service Corps.

Landownership Management—33% Reduction

A 33 percent decrease in funding to Landownership Management has impaired the agency's ability to respond to demands that are growing in number and complexity in the lands special use permits program, which supports critical projects involving energy pipelines, geothermal, electric transmission, hydropower, telecommunication infrastructure, including cellular towers and traditional line service and broadband facilities.

These special use projects provide community development and growth that directly supports job creation. Reductions in funding have limited the agency's ability to keep up with title claims and encroachments, each of which can cost the agency thousands of dollars to investigate and resolve, or can lead to even more costly lawsuits.

Wildlife & Fisheries Habitat Management—18% Reduction

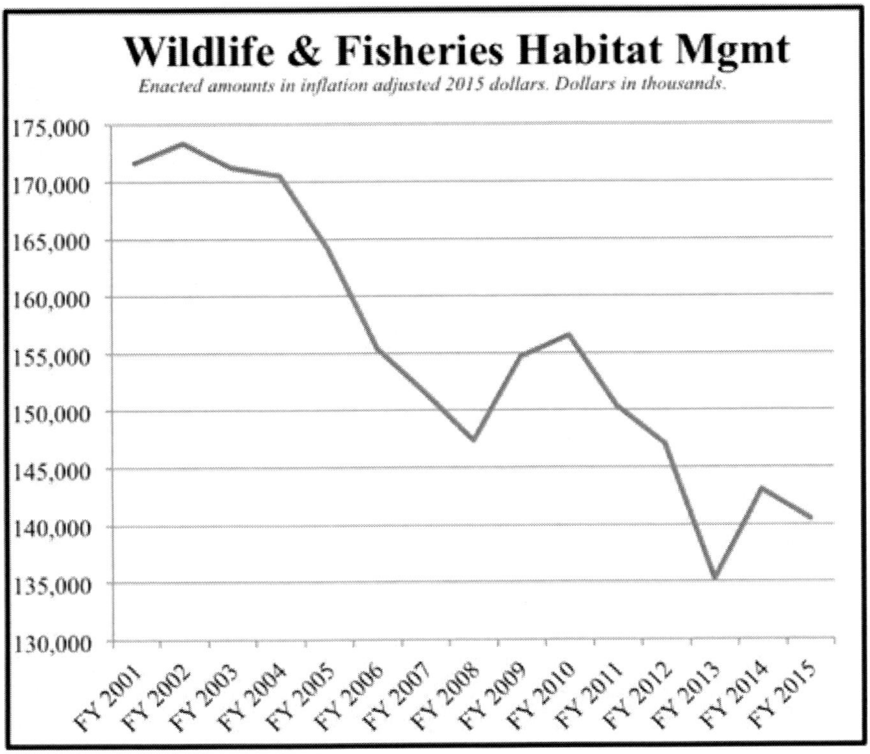

The reduction in funding to Wildlife and Fisheries compromises, for example, recovery efforts for threatened and endangered species (TES). The Forest Service has been unable to fulfill all of the required monitoring associated with previous Endangered Species Act (ESA) consultation commitments, reducing the agency's ability to implement projects in the future and jeopardizing current projects because ESA consultation requirements cannot be met. The agency has had to forego many projects critical to TES recovery and conservation efforts and offsetting the impacts of climate change.

The reduction in operating funds has limited the agency's ability to support existing partnerships. These partnerships often bring in as much as four-to-one return in partner contributions. This has resulted in the loss of significant dollars that could have been leveraged and has further reduced the agency's impact on key restoration objectives.

Land Management Planning—64% Reduction

Reductions in Land Management Planning have had a significant impact on the Forest Service's ability to reduce the backlog of forest plans requiring revision. The National Forest Management Act requires that each unit of the NFS have a Land Management Plan (LMP) that is formally revised every 10 to 15 years to address changing conditions and new information related to natural resources, management goals, and public use. The agency has only been able to meet this revision requirement on about 46 percent of its 125 LMPs.

The updated LMPs are essential for providing current, broad guidance for identifying, prioritizing, and implementing the programs and projects that move an NFS unit towards achieving desired conditions and achieving agency objectives. Reduced funding has had a significant effect on our ability to engage with the public and partners to address management issues and opportunities that have emerged since the original plans were developed. These efforts are essential for garnering public support and reducing appeals and litigations, which impacts our ability to implement key restoration efforts and increases implementation *costs.*

Inventory & Monitoring—35% Reduction

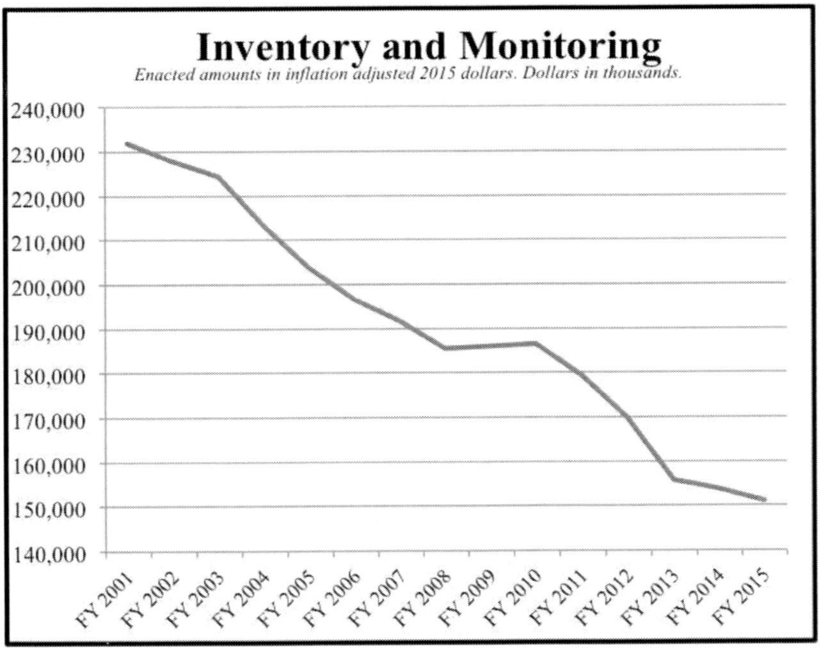

Reductions in Inventory and Monitoring have reduced the agency's ability to collect essential inventory and monitoring information, which has further delayed our ability to revise forest management plans. This reduced funding has hampered the Forest Service's ability to plan and execute projects for adapting and mitigating the effects of changing climate conditions, including completing watershed condition assessments, developing strategies for addressing needs for specific wildlife species and roadless area evaluations, as well as ensuring abundant clean water, providing recreation opportunities, restoring and maintaining forest and rangeland ecosystems, and improving priority watersheds across larger landscapes. Without current and adequate planning and monitoring, our ability to effectively deliver restoration treatments, recreation and special use permitting, and other economic activities on NFS lands has been adversely affected.

CONCLUSION

As documented in this report, the rising cost of fire suppression coupled with the current budgeting model is significantly impacting all non-fire program and staff areas.

The dramatic underlying shift of funding and human capacity from non-fire programs to support fire programs has real implications on the ground, including for restoration work that would help prevent catastrophic fires, protect watersheds that provide clean drinking water to tens of millions of people, protect irreplaceable cultural resources, and provide the infrastructure and programming that supports the $646 billion outdoor recreation economy and jobs and economic growth in hundreds of rural communities.

To solve this problem, we must change the way we pay for wildfire. Instead of treating catastrophic wildfires as a normal agency expense, we must treat them more like other natural disasters, such as tornadoes or hurricanes. And any solution must confront both parts of the funding quandary: it must limit or reverse the runaway growth of firefighting costs, and it must address the compounding disruption of fire transfers.

Bipartisan legislation that offers a more rational approach to funding wildfire, the Wildfire Disaster Funding Act, has already been introduced in the House and Senate. It is mirrored by a similar option in the President's 2016 Budget. This proposal provides a fiscally responsible mechanism to treat wildfires more like other natural disasters, end transfers, and partially replenish agency capacity to restore resilient forests and protect against future fire outbreaks.

USDA and the Forest Service look forward to working with Congress to take action to address the growth of fire costs that is crippling the agency's ability to conserve the nation's forests and grasslands and to provide the multiple uses and values for which the agency was created.

End Notes

[1] Preparedness, Suppression, FLAME, and related programs.

[2] This projection was developed by Forest Service researchers and is based on similar methodologies currently used for suppression cost expenditures that are required within FLAME Act provisions.

[3] These charts start at 2001 rather than 1995 because many of the programs (BLIs) have changed over the years, and FY 2001 to FY 20015 represents a time period over which the following programs remained consistent and therefore provide for an analogous comparison year over year.

[4] Appropriation numbers were adjusted for inflation to constant FY 2015 dollars. The deflators are from the Office of Budget and Management Fiscal Year 2015 Historical Tables, Table 10.1 (http://www.whitehouse.gov/sites/default/files/omb/budget/fy2015/assets/hist.pdf).

In: Wildfire Suppression
Editor: Neal Jenkins

ISBN: 978-1-63483-773-6

Chapter 3

WILDFIRE MANAGEMENT: FEDERAL FUNDING AND RELATED STATISTICS*

Katie Hoover and Kelsi Bracmort

SUMMARY

Wildfires can have beneficial and harmful impacts on ecosystems (e.g., by reducing fuel loads, or by damaging communities and timber resources). These impacts are generally measured and discussed based on the priorities of humans in these ecosystems. Federal resources are typically deployed during wildfire season—an annual occurrence of intense wildfire activity typically running from summer into fall—to help manage wildfires and potentially minimize some of the impacts, including the loss of life and property. Approximately 3.6 million acres burned during the 2014 wildfire season, which was less than the 4.3 million acres burned in 2014 and the fewest acres burned since 2010. Acreage burned during the 2014 wildfire season was less than half of the 9.3 million acres burned in 2012, which was the third-largest acreage burned annually since 1960.

Federal funding for wildfire management is provided in the Interior, Environment, and Related Agencies appropriations bill. The bill funds wildfire management at the Forest Service and the Department of the Interior, which are the two principal entities tasked with federal wildfire

* This is an edited, reformatted and augmented version of a Congressional Research Service publication R43077, prepared for Members and Committees of Congress, dated February 4, 2015.

management. Federal wildfire response activities involve preparedness, suppression, fuel reduction, site rehabilitation, and more. More than $3.4 billion was appropriated for wildfire management in FY2015. The total wildfire management appropriations for 2014 were more than $3.9 billion and included $600 million for the Forest Service to reimburse wildfire transfers that occurred in FY2013.

Wildfire funding is of concern to many. This is primarily due to interest in the effectiveness of annual wildfire appropriations (e.g., whether the funding is allowing agencies to meet wildfire management targets), the amount of appropriations dedicated to certain wildfire management activities (e.g., whether enough funding is being provided for hazardous fuel reduction), and requests from the agencies for additional appropriations during severe fire activity (e.g., why more funding is needed, and what non-fire programs could be impacted if the agencies fail to receive the additional funding). Congress has provided funding for wildfire management additional to what was granted in the Interior appropriations bill—usually for wildfire suppression—for 7 of the last 10 years, including FY2013 and FY2014. This fact may cause some to question how wildfire funding is structured and wildfire suppression funding estimation methods, among other things.

Proposals to create alternative mechanisms for funding wildfire suppression have been introduced in the 114th Congress (H.R. 167 and S. 235). Similar bills were introduced in the 113th Congress as well as proposed by the Administration for FY2015 and FY2016. The proposals would fund wildfire suppression for a select group of fires—theoretically the most extreme fires—by creating a new adjustment to the statutory discretionary spending limits.

This report provides wildfire management appropriations data for FY2011-FY2015 and information on the President's FY2016 budget request for wildland fire management. The report also provides general wildfire management statistics (e.g., numbers of wildfires, acres burned, and firefighter personnel).

INTRODUCTION

In 2014, wildfire burned approximately 3.6 million acres.[1] This was below the 10-year average of 6.9 million acres and was the least amount of acreage burned since 2010. The 2014 wildfire season also burned less than half of the 9.3 million acres burned during the 2012 wildfire season. The acreage burned in 2012 was the third-largest annual acreage burned, based on the reporting of fire statistics from 1960 to the present.[2] Some predict conditions that could lead to more severe wildfire activity potentially occurring in 2015 in certain regions, and possibly an earlier start for the 2015 wildfire season in certain regions due to ongoing drought conditions in the western half of the nation.[3] Wildfire (or wildland fire),[4] an unplanned and unwanted fire, is a concern to some because it can lead to loss of human life,[5] damage communities and timber resources,[6] and impact soils, watersheds, water quality, and wildlife.[7] On the other hand, wildfire can also reduce fuel loads, increase ecosystem health, and restore fire-adapted ecosystems.

There are several ongoing concerns regarding wildfire management. These include the total federal costs of wildfire management, the strategies and resources used for wildfire management, and the impact of wildfire on both the quality of life and the economy of communities surrounding wildfire activity. Many of these issues are deliberated by Congress, with annual wildfire management appropriations being one indicator of how Congress prioritizes and addresses certain wildfire management concerns. More than $3.5 billion was appropriated for wildfire management in FY2015. In FY2014, more than $3.9 billion was appropriated, including $600 million for the Forest Service to reimburse wildfire transfers that occurred in FY2013.

This report provides some general wildfire statistics, and it then briefly describes wildfire management appropriations and related issues.[8]

WHAT IS WILDFIRE MANAGEMENT?

Wildfire management is a series of coordinated activities undertaken by federal, state, and local authorities to resolve wildfire events when they occur. Resolution may involve immediate and aggressive measures to quickly suppress a wildfire (e.g., personnel and large air tanker response for a wildfire moving quickly toward a populated area), or may involve immediate but less intense measures (e.g., monitoring a grassland wildfire where there is no immediate threat to humans and weather conditions are favorable for natural suppression of the fire within a short time period).

Wildfire management activities are generally categorized into four areas: preparedness, suppression, fuel reduction, and site rehabilitation.

- *Preparedness* is any activity that leads to a safe, efficient, and cost-effective fire management program, and includes the range of tasks necessary to build, sustain, and improve the capability to protect against, respond to, and recover from domestic incidents.[9]

- *Suppression* is all of the work associated with extinguishing or confining a fire.[10]
- *Fuel reduction* is manipulation (including combustion) or removal of fuels to reduce the likelihood of ignition and/or to lessen potential damage and resistance to control.[11]
- *Site rehabilitation* includes efforts undertaken, generally within three years of a wildfire, to repair or improve fire-damaged lands unlikely to recover to a management-approved condition within a specified time frame, or actions taken to repair or replace minor facilities damaged by fire.[12]

A number of federal, state, and local agencies can and do respond to wildfire.[13] Sometimes more than one agency may respond, depending on where the fire occurs, the potential threats, and the expertise required. In these cases, the National Interagency Fire Center (NIFC)[14] coordinates the national mobilization of resources for wildfire and other incidents throughout the United States.[15]

WILDFIRE MANAGEMENT STATISTICS

Wildfire statistics help to illustrate past U.S. wildfire activity. Most of the statistics provided in this section are national in scope. NIFC data (*Figure 1*) indicate that the number of wildfires reported has stayed about the same during the 1983 to 2014 period, whereas the number of acres burned has increased steadily. *Table 1* gives the corresponding number of wildfires and acres burned for 2010 through 2014.

Table 1. Total Number of Wildfires and Acres Burned, 2010-2014

	2010	2011	2012	2013	2014
No. of Fires (in thousands)	72.0	74.1	67.8	47.6	63.2
Acres Burned (in millions)	3.4	8.7	9.3	4.3	3.6

Source: NIFC, Wildland fires and acres.

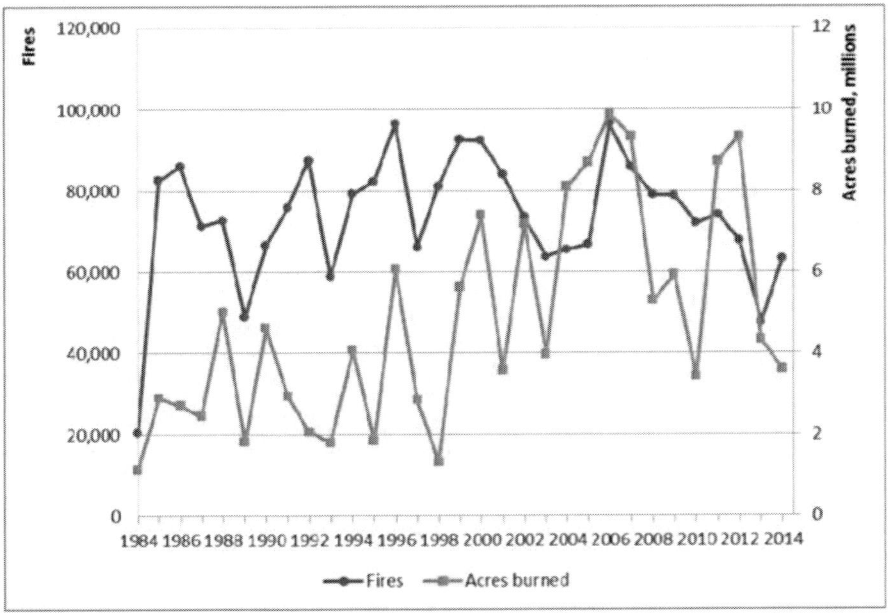

Source: National Interagency Fire Center (NIFC).

Figure 1.Annual Trends in Wildfires and Acres Burned, 1984-2014.

Table 2. Total Hazardous Fuels Acres Treated by Forest Service (FS) and Department of the Interior (DOI), FY2010-FY2014 (in millions)

	FY2010	FY2011	FY2012	FY2013	FY2014
FS	3.26	2.77	2.62	2.63	2.54
DOI	1.30	0.99	1.00	0.64	0.90

Source: Emails from FS on April 25, 2013, February 24, 2014, and January 20, 2015. DOI data are from the National Fire Plan Operations Reporting System, agency budget justifications, and emails from DOI on February 24, 2014, and January 26, 2015.

NIFC reports that in 2014, there were 63,212 wildfires and 3.6 million acres burned (data compiled from federal and state agencies).[16] More fires burned in 2014 than 2013, although fewer acres burned. More detailed statistics on the 2014 fire season are not yet available. However, of the 47,579 wildfires that burned over 4.3 million acres in 2013, less than 2% were classified as large or significant wildfires.[17] Approximately 71% of the number of wildfires reported in 2013 occurred on state or other land, but approximately 68% of the acres burned were on federal land (2.9 million acres). The Forest Service (FS) had the largest percentage (15%) of wildfires in 2013, in terms of number of fires occurring on federal land. The FS and Bureau of Land Management (BLM) had the largest percentage of acres burned on federal land, at 32% (0.9 million acres) and 23% (0.7 million acres), respectively.

Both the FS and Department of the Interior (DOI) keep statistics on hazardous fuels treatment: a measure that may help to mitigate the damaging effects of wildfire activity.[18] Hazardous fuel treatment includes prescribed burning, thinning, and more. Acres treated for FS and DOI lands are provided in *Table 2*.

Wildfires can have a devastating impact, especially for those communities adjacent to wildfire activity. Therefore, statistics showing the level of destruction a wildfire causes can be useful. Some more easily quantifiable measures include acres burned, lives lost (firefighters and civilians), structures destroyed, and suppression costs for a specific wildfire. There is no single source for this type of "destructive" data. Multiple federal and state organizations collect different data. Therefore, the data can usually be found, but in a piecemeal fashion. Firefighter personnel data for the FS and DOI, firefighter fatalities, and structures burned are provided in *Table 3*.

Table 3. FS and DOI Firefighting Personnel, Fatalities, and Structures Burned, FY2010-FY2014

	FY2010	FY2011	FY2012	FY2013	FY2014
Personnel					
FS Firefighters[a]	10,480	10,480	10,480	10,480	10,000
DOI Firefighters[b]	3,664	3,664	3,447	3,450	3,450
Losses					
Firefighter Fatalities[c]	8	12	15	34	10
Structures Burned[d]	788	5,246	4,244	2,135	1,953

Source: Agency budget justifications and agency emails. Firefighter fatalities data from NIFC, Historical Wildland Firefighter Fatality Reports and NWCG, Fatalities, Entrapments and Serious Accident Summary Safety Gram. Structures burned year-end totals from the NIFC Wildland Fire Summary and Statistics Annual Report 2013.

[a] FS firefighter data include hot shot crews, other firefighters, smoke jumpers, and prevention technicians.

[b] DOI firefighter data include firefighters, smoke jumpers, and Type 1 crews. Although provided in the DOI justification, fire program and support staff are omitted in this table from the DOI firefighter total. DOI reports that the personnel data do not include administratively determined (AD) emergency firefighters who are hired for short durations, often for a single incident. AD firefighter data can fluctuate significantly based on need. Source: Email from DOI on May 2, 2013 and January 26, 2015.

[c] Corresponds with calendar year.

[d] Structures include primary residences, commercial buildings, outbuildings, and seasonal dwellings. Corresponds with calendar year.

WILDFIRE MANAGEMENT APPROPRIATIONS

Wildfire management has been a responsibility of the federal government for over a century. Appropriations for wildfire management are funded in Interior, Environment, and Related Agencies appropriations bills, with both the FS and DOI receiving appropriations to manage wildfires. The FS carries out wildfire response in national forests and national grasslands. DOI manages the wildfire response in national parks, wildlife refuges and preserves, Indian reservations, and other public lands. Wildfire management funding for DOI is provided to the department, which then allocates the funding to the Office of Wildland Fire and four agencies— Bureau of Land Management, Bureau of Indian Affairs, National Park Service, and the U.S. Fish and Wildlife Service. Wildfire management appropriations began to increase after FY2000— partly in response to the severe 2000 fire season—and peaked at $4.5 billion in FY2008 (*Figure 2*).[19] Over the last five years (FY2011-FY2015), wildfire management appropriations for both the FS and DOI combined have averaged $3.3 billion (*Figure 3*).[20]

Forest Service (FS)Wildfire Management Appropriations

A majority of wildfire management appropriations go to the FS. From FY2011 to FY2015, the FS received on average 76% of the total wildfire management appropriations. In addition, FS wildfire management appropriations constitute a large portion of the FS overall discretionary funding total. For example, in FY2015 FS wildfire management appropriations are approximately 52% of the total discretionary funds appropriated to the FS (*Figure 4*).

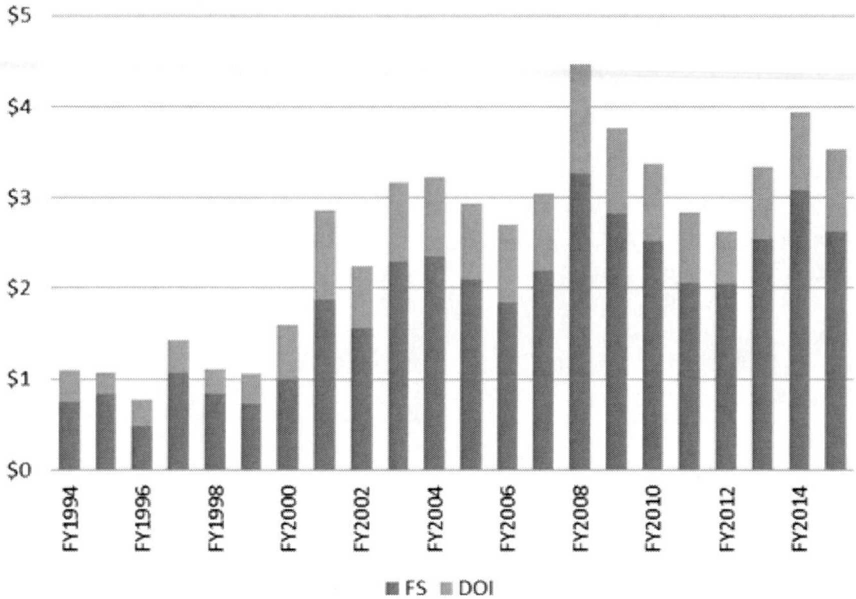

Source: CRS. Data compiled from detailed funding tables prepared by the House Committee on Appropriations for FY2005 through FY2014 and CRS Report RL33990, Federal Funding for Wildfire Control and Management.

Notes: Total wildfire management appropriations include appropriations to the Wildland Fire Management (WFM) account, the Federal Land Assistance, Management, and Enhancement Act (FLAME) account, and additional funding for wildfire.

Figure 2. Total Wildfire Management Appropriations, FY1994-FY2015. (billions of dollars)

The Interior appropriations law contains two accounts for wildfire for both the FS and DOI: the Wildland Fire Management (WFM) account and Federal Land Assistance, Management, and Enhancement Act (FLAME) account. Each agency's WFM appropriation is distributed among various activities and subactivities.[21] Two activities—preparedness and suppression—receive the bulk of the WFM appropriations, followed by hazardous fuels. FLAME is a reserve fund for wildfire suppression that requires certain conditions to be met in order to transfer funding from FLAME to the WFM account. DOI reports that the FLAME fund is used for "the most severe, complex, and threatening fires, and to serve as a contingency reserve."[22] *Table 4* contains annual total wildfire management appropriations from FY2006 to FY2015. Both the WFM account and the FLAME account are explained in more detail in the following sections of this report, as is additional funding.[23]

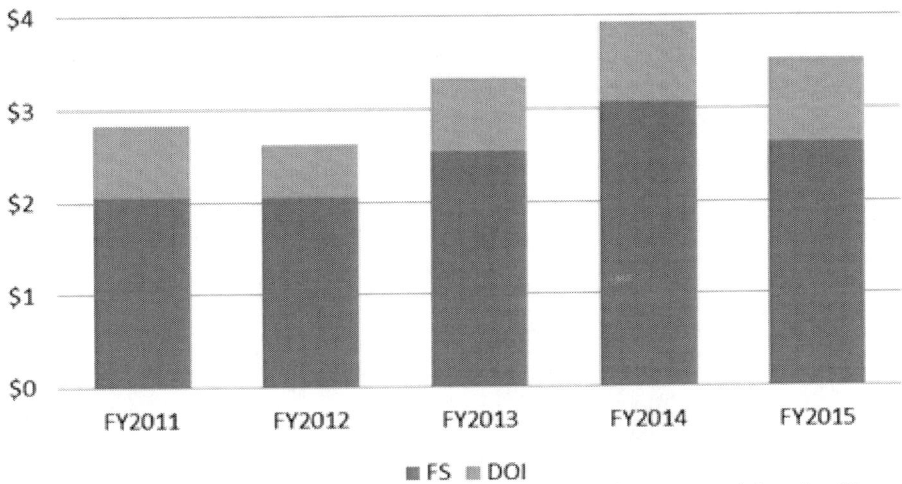

Source: CRS. Data compiled from detailed funding tables prepared by the House Committee on Appropriations.

Notes: Total wildfire management appropriations include appropriations to the WFM account, the FLAME account, and additional funding for wildfire. Figure 3 includes additional funding to repay previous-year wildfire accounts for FY2013 and FY2014 for the fiscal year in which appropriated (e.g., $395.4 million appropriated in 2013 to repay 2012 FS and DOI wildfire suppression accounts is reflected in the FY2013 column; $628.5 million appropriated in 2014 to repay 2013 FS and DOI wildfire accounts is reflected in the FY2014 column).

Figure 3. Total Wildfire Management Appropriations for FY2011-FY2015. (billions of dollars)

Preparedness

Appropriations for preparedness are used to support efforts that assist with fire prevention and detection, equipment, training, and baseline personnel. The FS uses preparedness funds to establish and maintain Web-based decision support tools, manage aviation assets, and offer predictive service analysis, among other things. DOI uses preparedness funds to protect people and property from wildfire with the use of fire management plans, cooperative agreements, infrastructure support, and more.

For the last five years, the preparedness activity for both the FS and DOI has received the largest or second-largest share of total WFM account appropriations. There was an increase of close to 50% in FS preparedness funding from FY2011 ($673.7 million) to FY2012 ($1,004.4 million). This was mostly due to a restructuring of the preparedness and suppression activities, which included shifting aviation charges and other charges to the preparedness activity. The funding level for the DOI preparedness activity saw minor fluctuations from FY2011 through FY2014 but increased 10% with the FY2015 appropriation ($319.0 million).

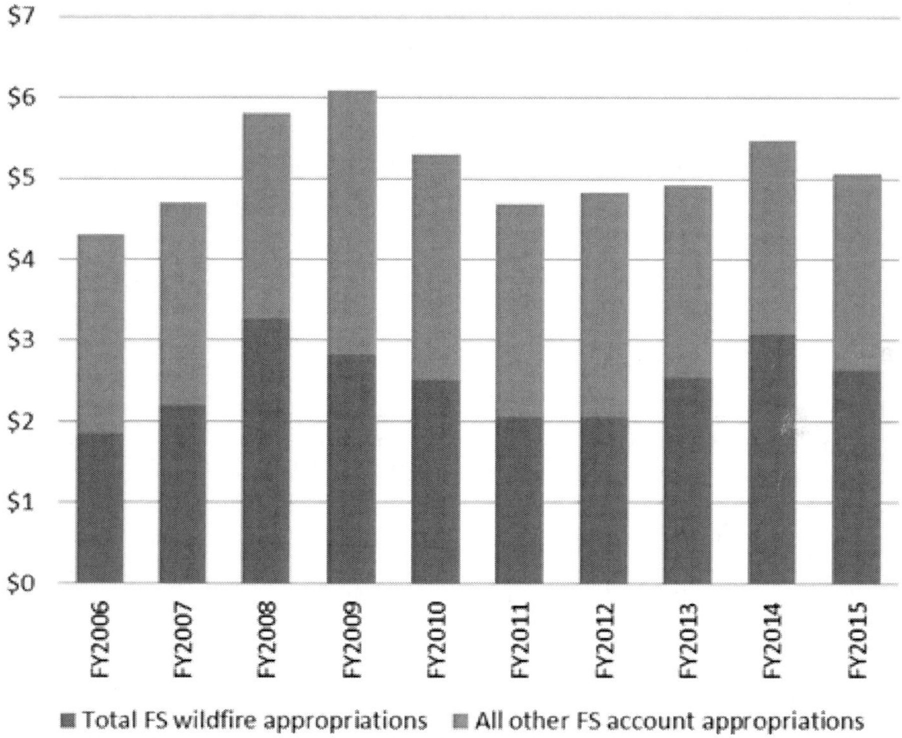

Source: CRS. Data compiled from detailed funding tables prepared by the House Committee on Appropriations.

Notes: Total wildfire appropriations include appropriations to the WFM account, the FLAME account, and additional funding for wildfire. All other FS account appropriations include appropriations for Forest Rangeland and Research, State and Private Forestry, National Forest System, Capital Improvement and Maintenance, Land Acquisition, and Other.

Figure 4. Total FS Wildfire Appropriations Compared to Other FS Account Appropriations for FY2006-FY2015. (billions of dollars)

Table 4. Total Wildfire Enacted Appropriations, FY2006-FY2015 (millions of dollars)

Agency/Account/ Activity	FY2006	FY2007	FY2008	FY2009	FY2010	FY2011	FY2012	FY2013[a]	FY2014	FY2015
Forest Service										
WFM Account[b]	1,746.1	1,823.6	1,943.5	2,131.6	2,178.7	2,168.0	1,734.9	1,868.9	2,162.3	2,333.3
Preparedness	660.7	665.4	665.8	675.0	675.0	673.7	1,004.4	948.7	1,057.6	1,145.8
Suppression	690.2	741.5	845.6	993.9	997.5	995.5	538.2	509.8	680.5	708.0
Hazardous Fuels	280.1	301.3	310.1	328.1	350.3	349.6	317.1	301.1	306.5	361.7
Rehabilitation[c]	6.1	6.2	10.8	11.5	11.6	11.5	—	—	—	—
FLAME Account	—	—	—	—	413.0	290.4	315.4	299.0	315.0	303.1
Additional Appropriations	100.0	370.0	1,326.0	700.0	—	—	—	379.9	600.0	—
Rescission	—	—	—	—	—	-400.0	—	—	—	—
Use of Prior Year Funds	—	—	—	—	-75.0	—	—	—	—	—
FS Total	1,846.1	2,193.6	3,269.5	2,831.6	2,516.7	2,058.5	2,050.2	2,547.6	3,077.3	2,636.4
Department of the Interior										
WFM Account	755.3	758.4	808.1	859.5	919.9	918.1	755.2	687.5	741.0	804.8
Preparedness	268.8	274.9	276.5	281.8	290.5	290.5	276.5	264.8	281.9	319.0
Suppression	230.7	249.2	289.8	335.2	383.8	399.0	270.5	261.2	285.9	291.7
Hazardous Fuels	208.1	199.8	199.6	203.1	206.2	183.3	183.0	137.7	145.0	164.0
Reduction Burned Area	24.1	22.8	24.2	20.3	20.3	33.2	13.0	12.3	16.0	18.0
Rehabilitation										

Table 4. (Continued)

Agency/Account/ Activity	FY2006	FY2007	FY2008	FY2009	FY2010	FY2011	FY2012	FY2013[a]	FY2014	FY2015
FLAME Account	—	—	—	—	61.0	60.9	91.9	91.7	92.0	92.0
Additional Appropriations	100.0	95.0	384.0	65.0	—	—	—	23.0	36.0	—
Rescission	—	—	—	—	—	-200.0	-82.0	-7.5	-7.5	—
Use of emergency suppression funds	—	—	—	—	—	—	-189.6	—	—	—
Use of Prior Year Funds	—	—	—	—	-125.0	—	—	—	—	—
DOI Total	855.3	853.4	1,192.1	924.5	855.9	778.9	575.4	794.7	861.5	896.8
Total (FS & DOI)										
Total, WFM Account	2,505.7	2,596.1.0	2,751.5	2,991.1	3,086.1	2,886.1	2,490.0	2,556.4	2,903.3	3,138.1
Total, FLAME Account	—	—	—	—	474.0	351.3	407.2	390.6	407.0	395.1
Total, Additional	200.0	465.0	1,710.0	765.0	—	—	—	402.9	628.5	—
Appropriations[d]										
Total[e]	2,705.7	3,061.1	4,470.3	3,762.9	3,379.6	2,844.3	2,638.5	3,342.4	3,938.8	3,533.1

Source: Information in this table is derived from detailed funding tables prepared by the House Committee on Appropriations and agency FY2015 budget justifications.

Notes: Figures generally reflect supplemental appropriations and rescissions, but do not reflect scorekeeping adjustments. Total equates to the summation of the WFM account, the FLAME account, additional appropriations, rescissions, and use of prior year funds. Wildfire appropriations for DOI used to go directly to BLM, but since 2009 appropriations have gone to DOI to be allocated among the agencies.

[a] FY2013 reflects post-sequester appropriations.

[b] Does not include state fire assistance or volunteer fire assistance funding under the FS State and Private Forestry (SPF) account for those years when appropriations were provided under the SPF account. Prior to FY2014, appropriations for state fire assistance and volunteer fire were provided to both the SPF account and the WFM account. In 2014, appropriations for state fire assistance and volunteer fire assistance for the SPF account were shifted to state fire assistance and volunteer fire assistance for the WFM account. For both the FS and DOI, activity values provided in this table for the WFM account when combined will not equate to the total WFM account value because there are other activities not included in the table.

[c] There is no specific appropriation for this activity after FY2011. The FS proposed to terminate this line item in FY2012, shifting the responsibilities for this subactivity to the National Forest System account, specifically the Integrated Resource Restoration budget line item.

[d] Total additional appropriations do not include rescissions, use of emergency suppression funds, or use of prior year funds.

[e] Total reflects final wildfire appropriation to FS and DOI for WFM, FLAME, and additional appropriations as well as any rescissions, use of emergency funds, or use of prior year funds.

Suppression

Funds for suppression typically are provided through two accounts in annual Interior appropriation laws: the WFM account and the FLAME account. In addition, other appropriation laws, such as supplemental laws, sometimes provide additional suppression funds for these accounts. This section of the report focuses on appropriations to the suppression activity under the WFM account. FLAME and additional suppression appropriations are discussed in the FLAME section and the additional funding section.

WFM suppression appropriations are used primarily for wildfire response. The FS reports that funding from this subaccount is used to "efficiently suppress wildland fires on or threatening National Forest System lands and other Federal and non-Federal lands under fire protection agreements."[24] Some items covered by the FS suppression subaccount are firefighter salaries, aviation asset operations, incident support function, and suppression resources for DOI incidents on a reciprocal non-reimbursement basis. The FS suppression subaccount also covers personnel and resources for the Burned Area Emergency Response (BAER) program.[25] DOI reports that wildfire response activities range from "intensive suppression when wildfires on public lands threaten communities, high value resources, or critical ecosystems, to monitoring wildfires in areas where burning accomplishes resource benefits or where it is too dangerous to place firefighters."[26] Items covered by the DOI suppression subaccount include select personnel expenses above what is covered by the preparedness subaccount, temporary emergency firefighters, and aircraft flight operations and support. Both the FS and DOI typically request suppression funds based on the 10-year suppression average. The 10-year suppression average is more fully explored in the "Related Issues" section.

The suppression subaccount received the greatest level of funding compared to all other WFM activities for the FS from FY2006 to FY2011 and for DOI from FY2008 to FY2011. However, from FY2012 to FY2015, appropriations provided for the FS suppression activity were second to the appropriations provided for the preparedness activity due to budget restructuring that occurred in FY2012.[27] From FY2012 through FY2014, appropriations provided for the DOI suppression activity and the DOI preparedness activity were roughly equivalent, within $5 million of each other. However, the FY2015 appropriation for DOI suppression activity is $27 million below the FY2015 preparedness activity.

Hazardous Fuels

The Forest Service's hazardous fuels appropriation and DOI's hazardous fuels reduction appropriation are used to decrease fuel loads, or alter the arrangement of fuel loads, on federal lands to make fires less intense and more controllable. Appropriations for this subaccount are used for fuel reduction on federal lands and to treat high-priority areas in the wildland-urban interface (WUI).[28] The FS reports that this activity aims to "protect communities (and associated lives, property, and public infrastructure) and important assets which are at the highest risk from damaging wildfire, provide an environment where wildland fire can be managed more safely and effectively, and create healthy landscapes that help maintain our Nation's wealth and provide places for the American public to visit and enjoy nature."[29] Additionally, DOI reports that the hazardous fuel reduction program "removes or modifies vegetation to restore and maintain healthy, diverse ecosystems; reduces wildfire risks to communities and their values; reduces the risk of severe and potentially dangerous wildfire behavior; lessens post-wildfire damage; and limits the spread and proliferation of invasive species and detrimental pathogens."[30] Since FY2005, the hazardous fuels subaccount has received the third-largest share of WFM appropriations for both agencies.

FLAME

Congress established a FLAME account—under the Federal Land Assistance, Management, and Enhancement Act of 2009 (Title V of Division A of the FY2010 Department of the Interior, Environment, and Related Agencies Appropriations Act, P.L. 111-88)—to cover the costs of large or complex fires, when amounts provided in the FS and DOI WFM accounts for suppression and emergency response are exhausted. Both the Secretary of Agriculture and the Secretary of the Interior may transfer funds from their respective FLAME account into the respective WFM account, for suppression activities, upon a secretarial declaration. The declaration may be issued if the fire covers at least 300 acres or threatens lives, property, or resources, among other criteria. The conferees of the FY2010 Interior appropriations bill stated their intent that the funding in the FLAME account, together with appropriations to the WFM account, should fully fund anticipated wildfire suppression needs and prevent future borrowing of funds from non-fire programs.[31] DOI reports that 179 wildfires were eligible for FLAME funding in FY2013 (24 on DOI land and 155 on FS land).[32] The FS reports that the Secretary of Agriculture signed two declarations in FY2013. The first declaration was specifically for 31 wildfires, while the second declaration did not request funds for specific wildfires; rather, it requested "the remainder of available funds due to the exhaustion of funds within the suppression account."[33] *Table 5* contains FLAME account funds for FY2006 to FY2015.

Table 5. Wildfire Management Additional Funds and FLAME Account, FY2006-FY2015 (millions of dollars)

	FY2006	FY2007	FY2008	FY2009	FY2010	FY2011	FY2012	FY2013	FY2014	FY2015
FS	100.0	370.0	1,326.0	700.0	413.0	290.4	315.4	678.9	915.0	303.1
Additional appropriations	100.0	370.0	1,326.0	700.0	—	—	—	379.9	600.0	—
FLAME	—	—	—	—	413.0	290.4	315.4	299.0	315.0	303.6
DOI	100.0	95.0	384.0	65.0	61.0	60.9	91.9	107.2	120.5	92.0
Additional appropriations	100.0	95.0	384.0	65.0	—	—	—	15.5	28.5	—
FLAME	—	—	—	—	61.0	60.9	91.9	91.7	92.0	92.0
Total	200.0	465.0	1,710.0	765.0	474.0	351.3	407.3	786.1	1,035.5	395.1

Source: Annual agency budget justifications, conference agreements for select Interior appropriations laws, and detailed funding tables prepared by the House Committee on Appropriations.

Notes: FY2013 reflects post-sequester appropriations. CRS identified the additional appropriations in the table through an analysis of the Appropriations Committee tables accompanying the relevant appropriations bills for FY2008 to FY2015. Additional appropriations identified in the table were typically listed in the detailed committee funding tables as "supplemental, "additional," or "emergency." Some of the additional appropriations over the last seven years were formally designated so as to not be subject to certain statutory or procedural budget enforcement, as "emergency requirements" or for other purposes. For further information on emergency designations, see CRS Report R41564, Emergency Designation: Current Budget Rules and Procedures.

Supplemental Appropriations

Additional funds for WFM are typically appropriated through legislation that supplements the annual Interior appropriations law. In general, bill or report language specifies how the additional funds are to be used (e.g., for wildfire suppression, for fire transfer reimbursement, for emergency rehabilitation). At times additional appropriations were used to repay accounts from previous years. This shifts the timescale of when funding was needed, sometimes creating record-keeping challenges. Additionally, the severity of certain fire seasons (e.g., the 2012 fire season) fell beyond most expectations for typical emergency wildfire management appropriation needs. Additional funds, when combined with FLAME funds, have at times been at the same level or higher than some of the top-funded WFM subaccounts (e.g., preparedness, suppression). A breakdown of additional funds since FY2005 is provided in *Table 5*.

THE ADMINISTRATION'S FY2016 WILDFIRE MANAGEMENT REQUEST

The Administration requested a total of $4.2 billion for wildfire management for FY2016 ($3.21 billion for FS and $1.0 billion for DOI). This figure is an increase of $681.0 million from the FY2015 enacted level of $3.5 billion (see *Table 6*). However, much of this difference comes from a proposal to change wildfire suppression funding.

In FY2015 and again in FY2016, the Administration proposed a new funding mechanism for wildfire suppression activities, which included eliminating the FLAME reserve account. Therefore, the Administration did not request an appropriation to the FLAME reserve account either year (Congress did not adopt the FY2015 proposal). Currently, wildfire suppression is funded by appropriations that are subject to the statutory discretionary spending limits established in the Balanced Budget and Emergency Deficit Control Act of 1985.[34] The Administration's proposal would maintain this funding stream for most wildfire suppression operations, but it also would establish a new adjustment to the discretionary limits that could be used to fund the most extreme fires in emergency situations. In FY2016, the maximum amount of this adjustment would be $1.5 billion; this amount would increase each fiscal year until FY2022, when it would be $2.7 billion.[35] This adjustment would allow FS and DOI to request additional wildfire suppression funds—up to the maximum amount of the adjustment each fiscal year—that effectively would not be subject to the discretionary limits. The FS asserts that "fire suppression costs are skyrocketing and seriously jeopardizing [the FS's] ability to fund [the FS] natural resource mission."[36] The FS reports that the proposed strategy would "minimize the risk of fire transfers and provide greater stability and certainty of funding to other programs that support critical forest and rangeland management needs."[37] Similar legislative proposals have been introduced in the 114th Congress (H.R. 167 and S. 235).

Table 6. FY2015 Enacted Appropriations and FY2016 Budget Request for FS and DOI Wildfire Management (millions of dollars)

Agency/Account	FY2015 Appropriation	FY2016 Request
Forest Service		
WFM account	2,333.3	2,354.0
FLAME account	303.1	0
Wildfire Suppression Operations adjustment	—	854.6
Total, FS	2,636.4	3,208.6
DOI		
WFM account	804.8	805.5
FLAME account	92.0	0
Wildfire Suppression Operations adjustment	—	200.0
Total, DOI	896.8	1,005.5
FS and DOI		
WFM account	3,138.1	3,159.5
FLAME account	395.1	0
Wildfire Suppression Operations adjustment	—	1,054.6
Total, FS and DOI	3,533.1	4,214.1

Source: FY2015 enacted data from detailed funding tables prepared by the House Committee on Appropriations. FY2016 budget request from FS FY2016 Budget Justification, available from http://www.fs.fed.us/ about-agency/budget-performance, and DOI FY2016 Wildland Fire Management Budget Justification, available from http://www.doi.gov/budget/appropriations/2016/index.cfm.

Notes: The figure requested under the wildfire suppression operations adjustment would effectively not be subject to the discretionary spending limits.

RELATED ISSUES

Fire Transfers

The FS and DOI are authorized to transfer funds from other programs and accounts to fund wildfire suppression, typically when wildfire suppression accounts have been depleted. Transfer authority is granted in the Interior, Environment, and Related Agencies annual appropriations bill, specifically in the general provisions Section 102 for DOI and in the administrative provisions for the FS. The accounts from which funds were transferred have historically been reimbursed in the following fiscal year's appropriation act.

Such transfers of funds can have an impact on the ability of the agency to fulfill its other responsibilities.[38] Further, some surmise these fire transfers disrupt the congressional budget process.[39] When wildfire suppression funding is exhausted, Congress—sometimes outside of its regular budget schedule—faces a decision as to whether to reimburse the accounts from which funding was transferred. If a reimbursement decision is made after the end of the fiscal year when the transfers were made, it could create confusion about how much suppression funding is needed for the coming fiscal year during the appropriations process. Further, fire transfers may require the modification or suspension of a non-wildfire program if funding from that program is being transferred to wildfire suppression.[40]

Wildfire Suppression Funding Estimates

Wildfire management suppression appropriations are of considerable interest to Congress. This is partly because of the amount of funding allocated for wildfire management suppression, but also because of scrutiny about the accuracy of annual WFM suppression estimates provided to Congress. In some cases, wildfire management suppression estimates for a fiscal year provided by federal land management agencies have fallen short of what is needed. Additional funds may then have to be provided via additional appropriations, and funds may also have to be transferred from non-wildfire management suppression accounts, which may impact the performance of those activities.

For at least the last two decades, the FS and DOI have estimated annual wildfire suppression outyear costs by using a 10-year suppression average.[41] The FS reports the moving 10-year average calculation based on actual suppression obligations reported at the end of the fiscal year. Each end-of-year obligation is adjusted to constant dollars. Once an average is calculated, $350 million is subtracted to account for an FY2012 budget restructure in which funding was transferred from the suppression subaccount to the preparedness subaccount.[42]

FLAME requires the Secretaries of the Interior and Agriculture to submit to Congress a report at least three times throughout the year that contains estimates of anticipated wildfire suppression costs. Further, FLAME requires the FS and DOI to submit wildfire suppression cost estimates using the best available climate, weather, and other relevant data, and models and other analytical tools. This estimation is made using econometric models which include weather and climate variables.[43] For instance, the FS estimated with 90% confidence that wildfire suppression expenditures for FY2015 will range from $714 million to $1.5 billion, with a median forecast of $1.1 billion.[44] DOI estimated with 90% confidence that wildfire suppression expenditures for FY2015 will range from $247 million to $466 million, with a median forecast of $356 million.[45] Further, the September 2014 FLAME analysis indicates that from 2004 to 2013, annual observed FS wildfire suppression costs exceeded the FLAME forecasts six times, while the annual observed DOI wildfire suppression costs exceeded the FLAME forecasts seven times.[46]

Potential Wildfire Suppression Funding Alternatives

Wildfire suppression funding estimates depend on multiple factors (e.g., weather, fuel load, nearby dwellings, access to wildfire site), and numerous reasons have been given as to why suppression estimates have at times not accurately forecast the needed suppression appropriations, with estimates typically underestimating suppression spending. Wildfire suppression is complicated, and both the efficiency of resources used for wildfire suppression and the federal protocol for wildfire management have an impact on wildfire suppression funding.

Both Congress and the Administration have voiced concerns about rising wildfire suppression spending, and have explored multiple options that could limit requests for additional suppression appropriations.[47] Some cost-containment options include

- using an improved formula to better estimate wildfire suppression costs (e.g., a 5- year average, a weighted 10-year average),[48]

- using regression models to better estimate WFM suppression costs and non-biophysical policy remedies that might reduce wildfire suppression expenditures (e.g., guidelines about when it is appropriate not to aggressively suppress wildfires),[49]
- increasing hazardous fuels reduction projects,[50] and
- requiring state and local governments, and private landowners in the WUI, to improve current wildfire management strategies and bear more of the firefighting costs.[51]

Two proposals to create alternative mechanisms for funding wildfire suppression have been introduced to date in the 114th Congress (H.R. 167 and S. 235). These bills are similar to proposals from the 113th Congress (S. 1875 and H.R. 3992) as well as to proposals in the Administration's Wildfire Management Request in FY2015 and FY2016, discussed above.[52] H.R. 167 and S. 235 would fund wildfire suppression for a select group of fires—theoretically the most extreme fires—by creating a new cap adjustment to the statutory discretionary spending limits. The proposals also would treat larger wildfires as natural disasters. It is not clear if the proposed legislation would limit how much funding is spent on wildfire suppression.

End Notes

[1] Email from DOI on January 26, 2015.

[2] The largest fire year on record for acres burned is 2006 (9.8 million). The second- and third-largest fire years are 2007 (9.328 million) and 2012 (9.326 million). National Interagency Fire Center, Current Year-to-Date by State (2012), December 20, 2012; http://www.nifc.gov/fireInfo/fireInfo_stats_YTD2012.html, National Interagency Fire Center, Total Wildland Fires and Acres (1960-2009).

[3] National Weather Service, U.S. Seasonal Drought Outlook January 15 – April 30, January 15, 2015; National Interagency Fire Center, National Significant Wildland Fire Potential Outlook Period—January and February, through April 2015, January 1, 2015.

[4] The terms wildfire and wildland fire are often used interchangeably. However, each term has a distinct definition. The National Wildfire Coordinating Group (NWCG) defines wildland fire as any non-structure fire that occurs in the wildland, defined as an area in which development is essentially nonexistent, except for roads, railroads, power lines, and similar transportation facilities, and where structures, if any, are widely scattered. NWCG defines wildfire as an unplanned, unwanted wildland fire including unauthorized human-caused fires, escaped wildland fire use events, escaped prescribed fire projects, and all other wildland fires where the objective is to put the fire out. National Wildfire Coordinating Group, Glossary of Wildland Fire Terminology, July 2012. This report will use the term wildfire for both.

[5] An example is the 2013 loss of 19 firefighters of the Granite Mountain Interagency Hotshot Crew assigned to the Yarnell Hill fire. For more information, see CRS Report R43129, Wildfire Management: Hotshot Crews.

[6] For more information on wildfire damages to homes and resources, see CRS Report RL34517, Wildfire Damages to Homes and Resources: Understanding Causes and Reducing Losses.

[7] For more information, see CRS Report RL30755, Forest Fire/Wildfire Protection.

[8] Additional appropriations data and statistics prior to 2008 are available in CRS Report RL33990, Federal Funding for Wildfire Control and Management.

[9] National Wildfire Coordinating Group, Glossary of Wildland Fire Terminology, July 2012.

[10] Ibid.

[11] Ibid.

[12] U.S. Department of the Interior, Interagency Burned Area Emergency Response Guidebook, February 2006.

[13] For more information on federal assistance, see CRS Report R41858, Federal Assistance for Wildfire Response and Recovery.

[14] NIFC comprises eight agencies and organizations: the Bureau of Land Management, the Bureau of Indian Affairs, the U.S. Fish and Wildlife Service, the National Park Service, the U.S. Forest Service, the National Oceanic and Atmospheric Administration, the National Business Center, and the U.S. Fire Administration. Also, a National Association of State Foresters representative is located at NIFC. Additional information about the role of each agency is available in the NWCG Communicator's Guide for Wildland Fire Management: Fire Education, Prevention, and Mitigation Practices.

[15] The National Interagency Mobilization Guide identifies standard procedures that guide the operations of multi-agency logistical support activity.

[16] Email from Department of the Interior, January 26, 2015.

[17] Significant fires are defined as fires that are a minimum of 100 acres in timber fuel types, or 300 acres in grass and brush fuel types, or are managed by a Type 1 or 2 Incident Management Team. An incident management team consists of the incident commander and appropriate general and command staff personnel assigned to an incident. Type refers to resource capability. A Type 1 resource provides a greater overall capability due to power, size, capacity, etc., than would be found in a Type 2 resource.

[18] Additional information on hazardous fuels removal is available in the "Wildfire Management Appropriations" section of this report.

[19] For more information on wildfire management appropriations prior to FY2005, see CRS Report RL33990, Federal Funding for Wildfire Control and Management.

[20] This includes emergency funds (e.g., supplemental appropriations).

[21] Appropriations laws typically identify the appropriation for some purpose in general form. The supporting documentation accompanying appropriations bills, primarily conference reports, typically identify appropriations for activities specific to that purpose. Hereinafter, these purposes and activities below the WFM account level are referred to as "activities" for both the FS and DOI.

[22] U.S. Department of the Interior, Budget Justifications and Performance Information Fiscal Year 2015 Wildland Fire Management.

[23] Additional funding is any funding provided for wildfire management outside of the normal appropriations process.

[24] U.S. Forest Service, Fiscal Year 2014 Budget Justification, April 2013.

[25] The objective of the BAER program is to determine the need for and to prescribe and implement emergency treatments on federal lands to minimize threats to life or property

resulting from the effects of a fire or to stabilize and prevent unacceptable degradation to natural and cultural resources.

[26] U.S. Department of the Interior, Budget Justifications and Performance Information Fiscal Year 2015 Wildland Fire Management.

[27] A similar budget restructuring did not occur for DOI.

[28] For more information on WUI, see CRS Report RS21880, Wildfire Protection in the Wildland-Urban Interface.

[29] U.S. Forest Service, Fiscal Year 2015 Budget Justification, March 2014.

[30] U.S. Department of the Interior, Budget Justifications and Performance Information Fiscal Year 2015 Wildland Fire Management. In its FY2015 budget justification, DOI proposes to change the subactivity name from "Hazardous Fuels Reduction" to "Fuels Management."

[31] H.Rept. 111-316.

[32] U.S. Department of the Interior, Budget Justifications and Performance Information Fiscal Year 2015 Wildland Fire Management.

[33] U.S. Forest Service, Fiscal Year 2015 Budget Justification, March 2014.

[34] P.L. 99-177, Title II, 2 U.S.C. 900-922, as amended by the Budget Control Act of 2011. More information on the proposal is available as a CRS congressional distribution memorandum upon request.

[35] Office of Management and Budget, Analytical Perspectives, Budget of the United States Government, FY2016, February 2, 2015, pp. 131-132, http://www.whitehouse.gov/omb/budget/Analytical_Perspectives.

[36] U.S. Forest Service, The U.S. Forest Service—An Overview.

[37] See p. 251 of the FS Fiscal Year 2016 Budget Justification, available from http://www.fs.fed.us/about-agency/ budget-performance.

[38] For example, the FS Chief stated, "Each time the agency transfers money out of accounts to pay for fire suppression there are significant and lasting impacts across the entire Forest Service. Not only do these impacts affect the ability of the Forest Service to conduct stewardship work on national forests, they also affect our partners, local governments and Tribes." U.S. Congress, Senate Committee on Energy and Natural Resources, hearing to explore ways to improve federal wildland fire management, 113th Cong., June 4, 2013.

[39] The issue of fire transfers hindering the ability of the FS and DOI to achieve non-fire related goals is not new. For historical information on fire transfers, see CRS Report RL34772, Proposals to Merge the Forest Service and the Bureau of Land Management: Issues and Approaches, and U.S. General Accounting Office, Wildfire Suppression: Funding Transfers Cause Project Cancellations and Delays, Strained Relationships, and Management Disruptions, GAO-04-612, June 2004.

[40] For more information on FS fire funding transfers, see the FS FY2016 Budget Justification page 400, which contains a detailed account of FS fire transfers, specifically funds withdrawn for transfer for wildfire suppression in FY2013, and funds withdrawn for wildland fire suppression and not repaid from FY2002 to FY2014. Similar information was not available in the DOI FY2016 budget justification.

[41] Outyear is the fiscal year following the year covered in a budget.

[42] Wildfire 10-year suppression average explanation provided by the FS in emails on April 2, 2013, and February 27, 2014.

[43] Email from the FS May 14, 2013.

[44] DOI. Federal Land Assistance, Management and Enhancement (FLAME) Act Suppression Expenditures for Interior and Agriculture Agencies: FLAME September 2014 Forecasts for Fiscal Year 2015 (August 13, 2014).

[45] Ibid.

[46] Ibid. Annual observed FS wildfire suppression costs exceeded the FLAME Act May 2014 forecast in 2005, 2006, 2008, 2011, 2012, and 2013. Annual observed DOI wildfire suppression costs exceeded the FLAME ACT May 2014 forecast in 2005, 2006, 2007, 2008, 2011, 2012, and 2013.

[47] U.S. Congress, Senate Committee on Energy and Natural Resources, Wyden, Crapo Introduce Bipartisan Wildfire Funding Reform Legislation, 113th Cong., December 19, 2013; U.S. Department of Agriculture, "Obama Administration Outlines New Strategy to Better Protect Communities, Businesses and Public and Private Lands from the Threat of Wildland Fire Ahead of 2014 Fire Season," press release, April 9, 2014.

[48] U.S. General Accounting Office, Wildfire Suppression: Funding Transfers Cause Project Cancellations and Delays, Strained Relationships and Management Disruptions, GAO-04-612, June 2004.

[49] Karen L. Abt, Jeffrey P. Prestemon, and Krista M. Gebert, "Wildfire Suppression Cost Forecasts for the US Forest Service," Journal of Forestry, June 2009; Geoffrey H. Donovan, Jeffrey P. Prestemon, and Krista Gebert, "The Effect of Newspaper Coverage and Political Pressure on Wildfire Suppression Costs," Society and Natural Resources, vol. 24, no. 8 (2011).

[50] H.R. 1526, 113th Congress.

[51] Headwaters Economics, The Rising Cost of Wildfire Protection, June 2013.

[52] For more information on the Administration's proposal, see the FS Fiscal Year 2016 Budget Justification, p. 251, and the DOI Budget Justification and Performance Information Fiscal Year 2016, Wildland Fire Management.

In: Wildfire Suppression
Editor: Neal Jenkins

ISBN: 978-1-63483-773-6

Chapter 4

FIRE MANAGEMENT ASSISTANCE GRANTS: FREQUENTLY ASKED QUESTIONS*

Bruce R. Lindsay, Francis X. McCarthy, Katie Hoover and Kelsi Bracmort

SUMMARY

Section 420 of the Robert T. Stafford Disaster Relief and Emergency Assistance Act (P.L. 93-288) authorizes the President to "declare" a Fire Management Assistance Grant (FMAG). This authority has been delegated to the Federal Emergency Management Agency's Regional Administrators. Once issued, the FMAG declaration authorizes various forms of federal assistance such as the provision of equipment, personnel, and grants to state, local, and tribal governments for the control, management, and mitigation of any fire on certain public or private forest land or grassland that might become a major disaster.

The current FMAG system was established by regulation in October of 2001. Prior to that time, the program was known as the Fire Suppression Assistance Program. However, the program was administered in similar fashion with the FEMA Regional Administrators working with the requesting state and the "Principal Advisor," as well as

* This is an edited, reformatted and augmented version of a Congressional Research Service publication R43738, prepared for Members and Committees of Congress, dated September 25, 2014.

consulting with FEMA leadership prior to the announcement of Stafford Act assistance under Section 420.

This report answers frequently asked questions about FMAGs.

INTRODUCTION

Section 420 of the Robert T. Stafford Disaster Relief and Emergency Assistance Act (P.L. 93-288, hereinafter the Stafford Act)[1] authorizes the President to "declare" a Fire Management Assistance Grant (FMAG). These grants provide federal assistance for fire suppression activities. This authority has been delegated to the Federal Emergency Management Agency's (FEMA) Regional Administrators.[2] Once issued, the FMAG declaration authorizes various forms of federal assistance such as the provision of equipment, personnel, and grants to state, local, and tribal governments for the control, management, and mitigation of any fire on certain public or private forest land or grassland that might become a major disaster.[3] This federal assistance requires a cost-sharing component such that state, local, and tribal governments are responsible for 25% of the expenses.

The current FMAG system was established by regulation in October of 2001.[4] Prior to that time, the program was known as the Fire Suppression Assistance Program. The program, however, was administered in a similar fashion. Then, as now, the FEMA Regional Administrators worked with the requesting state, the "Principal Advisor," as well as consulting with FEMA leadership prior to the announcement of Stafford Act assistance under Section 420.

Declaration Process

How are FMAGs Requested?

FMAGS can be requested by a state when the governor determines that a fire is burning out of control and threatens to become a major disaster. At that point, a request for assistance can be submitted to FEMA. Typically, requests are submitted to the FEMA Regional Administrator. Requests can be submitted any time—day or night—and can be submitted verbally by telephone to expedite the process. Telephone requests must be followed by written confirmation within 14 days of the phone request.[5]

Can a Tribal Leader Request an FMAG Declaration?

Under the Sandy Recovery Improvement Act of 2013 (SRIA), tribes are now equivalent to states in their ability to request a major disaster or an emergency declaration from the President; however, tribes must continue to seek FMAG assistance through state requests.[6] During the pilot phase of direct tribal disaster declarations implementation, FEMA will evaluate the FMAG Program and determine if any changes are needed to FMAG declarations in this respect.[7]

It is helpful to note that tribal land holdings are largely federal land and, therefore, receive fire suppression support through the National Interagency Fire Center (NIFC). The NIFC supports the interagency "wildland" firefighting efforts on federal lands by the U.S. Forest Service, National Weather Service, National Park Service, Bureau of Indian Affairs (BIA), U.S. Fish and Wildlife Service and FEMA's U.S. Fire Administration.[8]

Unlike FMAGs, such support generally does not require tribes to reimburse the NIFC for firefighting costs (FMAGs requires the state to pay a 25% cost-share). In addition, tribes with their own fire suppression resources may receive reimbursement from BIA for their costs related to fire suppression on tribal lands.[9]

What Information Needs to Be Included in the FMAG Request?

The FMAG request should include factual data including any professional estimates to support the request. This information includes the size of the fire(s) in acres or square miles, the population of the community (or communities) threatened, the number of persons evacuated (if applicable), weather conditions, and the degree to which state and local resources are committed to this fire and other fires in federal, state, and/or local jurisdictions. The verbal request must be followed up with a completed "Request for Fire Management Assistance Declaration" (FEMA form 078-0-1) and the "Principal Advisor's Report" (FEMA form 078-0-2).[10]

How Is FMAG Assistance Determined?

The following criteria are used to evaluate wildfires and make a determination whether to issue an FMAG:

- The threat to lives and property including critical facilities, infrastructures, and watershed areas;
- the availability of state and local fire resources;

- high fire danger conditions based on nationally accepted indices such as the National Fire Danger Ratings System; and
- the potential economic impacts of the fire.

In addition, FEMA uses two types of fire cost thresholds to help determine if a state or tribal nation is eligible for fire assistance: (1) individual thresholds for a single fire, and (2) cumulative thresholds for multiple fires. Cumulative thresholds are applied to multiples fires burning simultaneously, or the accumulation of multiple fires in a single fire season. Threshold amounts vary by state (see *Table 1*).

Individual Fire Threshold

The formula for the individual fire threshold is the state population multiplied by 5%, which is then multiplied by $1.39.[11] In general, if that amount exceeds the state's individual fire threshold, the state is eligible for federal assistance.

For example, the state of Colorado's population is 5,029,196. The individual fire threshold formula for the state is 5,029,196 x 5% x $1.39 = $349,529. The state of Colorado would meet or exceed the individual fire threshold if it had a wildfire costing $349,529 or more in damages.

Cumulative Thresholds

The formula for the cumulative fire threshold for a given state is one of two amounts—$500,000 *or* the amount of that state's individual fire threshold multiplied by three, whichever is higher. Returning to the Colorado example, the sum of three individual fire thresholds equals $1,048,587. Since that amount is larger than $500,000, cumulative fire damages in Colorado.must meet or exceed $1,048,587 to be eligible for assistance. In contrast, the individual fire threshold for Alaska is $100,000, but the cumulative threshold is $500,000, not the sum of three individual fire thresholds ($300,000). For some states, such as Nebraska and West Virginia, the state population is high enough that the individual threshold exceeds $100,000. However, the cumulative threshold for these states is $500,000 because that number is still higher than the sum of three individual fire thresholds.

Table 1. Selected Examples of Individual and Cumulative Thresholds
FY2014 Individual and Cumulative Thresholds by State

State	Individual Threshold	Cumulative Threshold
Alaska	$100,000	$500,000
California	$2,589,150	$7,767,450
Colorado	$349,529	$1,048,587
Nebraska	$126,931	$500,000
West Virginia	$128,783	$500,000

Source: FEMA, "FY2014 Fire Cost Threshold," http://www.fema.gov/media-library-data/1389875349045-36f0ce0f02eed0357d 4fd151f7056c17/FY 14%20Fire%20Cost%20Threshold.pdf.

Can Denials for FMAG Assistance Be Appealed?

If FEMA denies the request for assistance, the state has one opportunity to appeal the denial. The appeal must be submitted in writing to the Regional Administrator no later than 30 days from the date of the denial letter. The appeal should contain any additional information that strengthens the original request for assistance. The Regional Administrator will review the appeal, prepare a recommendation, and forward the appeal package to the FEMA Headquarters Office. The FEMA Headquarters Office will notify the state of its determination in writing within 90 days of receipt of the appeal (or receipt of the additional requested information).

The state may request a time extension to submit the appeal. The request for an extension must be submitted in writing to the Regional Administrator no later than 30 days from the date of the denial letter. The request for an extension must include a justification for the need for an extension. The FEMA Headquarters Office will notify the state in writing whether the extension request is granted or denied.

Does an FMAG Exclude the Possibility of an Emergency or Major Disaster Declaration Under the Stafford Act?

No, an emergency or major disaster can be declared after an FMAG declaration has been issued. However, the emergency or major disaster declaration must be initiated by a separate request for assistance by the state or tribal government.[12]

How Many FMAGs Are Declared Each Year?

As shown in *Figure 1* the average number of FMAGs declared each year from FY2004 to FY2013 was 53. The highest number of FMAG declared in one year was 2011 (112 declarations). This surpassed the previous high of 94 FMAGs in 2006.

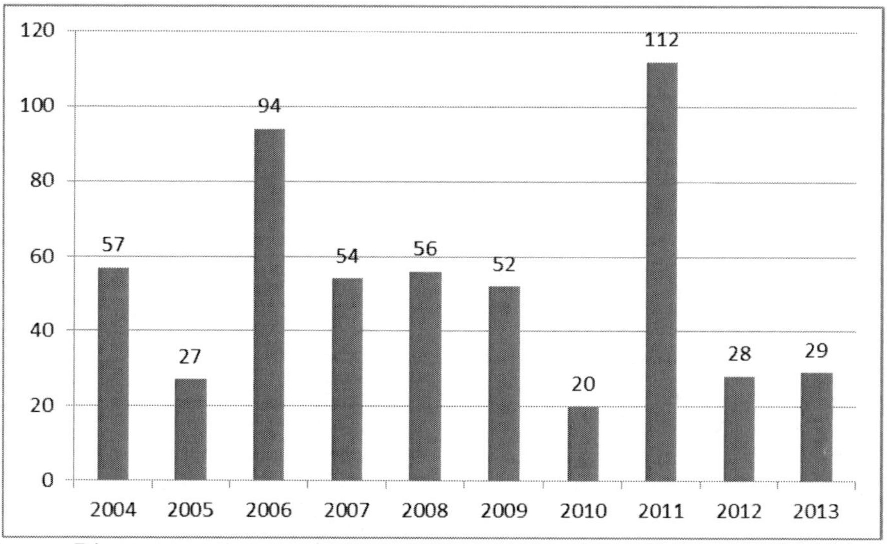

Source: Disaster cost data provided by FEMA, August, 2014. Declaration data can be located at https://www.fema.gov/disasters.

Figure 1. FMAG Declarations FY2004-FY2013

Funding[13]

How Are FMAGs Funded?

FMAG are funded through FEMA's Disaster Relief Fund (DRF), the main account FEMA uses to provide disaster assistance.[14] The DRF is a no-year account—unused funds from the previous fiscal year are carried over to the next fiscal year.

Funds in the DRF fall into two categories. The first category is for disaster relief costs associated with major disasters under the Stafford Act. This structure reflects the impact of the Budget Control Act (P.L. 112-25, BCA), which allows appropriations to cover the costs incurred as a result of major disasters to be paid through an "allowable adjustment" to the discretionary spending limits.[15] The second category is colloquially known as "base funding." Base funding includes activities not tied to major disasters under the Stafford Act. Base funding is scored as discretionary spending that counts against the discretionary spending limits. FMAGs are funded through the DRF's base category.

What Percentage of DRF Expenditures Are Provided for FMAGs?

Expenditures for FMAGs are relatively low compared to funding for emergencies, major disasters, and surge activities.[16] From FY2004 to FY2013, total obligations for the DRF were $98.7 billion. Of this total, only $682 million (0.7%) was for fire suppression activities.[17] Even for the single year of FY2011, in which obligations for FMAG's reached a historic high of $113.6 million, this amount only accounted for 1.6% of total DRF obligations. *Figure 2* shows total obligations for the period FY2004-FY2013 for major disasters, emergencies, FMAGs, and surge activities.[18]

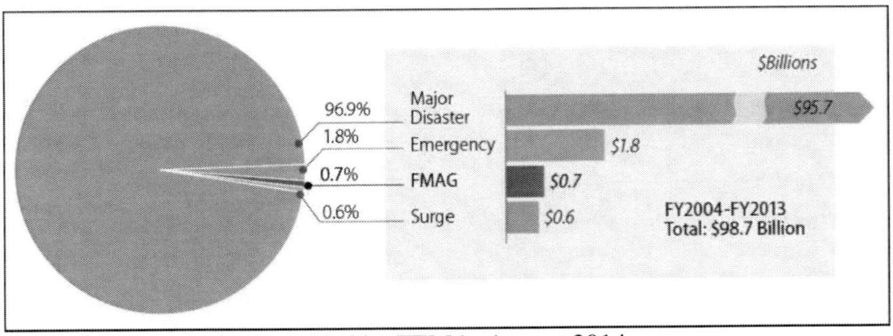

Source: Disaster cost data provided by FEMA, August 2014.

Figure 2. DRF Obligations By Type FY2004-FY2013.

What Is the Average FMAG Amount?

As demonstrated in *Figure 1*, from FY2004 to FY2013, 529 FMAGS were declared. Using the total obligations of $682 million during that time, the average obligation per FMAG was $1.29 million. However, there is a high level of variation for the total obligations associated with each declaration—some wildfires exceeded $10 million in total costs. As a result, the number of FMAG declarations in a given year is not an accurate predictor of total obligations from the DRF.

What Are the Budget Implications of FMAGs?

The budget implications of FMAGs are difficult to ascertain because on the one hand, an increase in FMAG declarations leads to more federal dollars being spent on wildfires. On the other hand, FMAGs are designed to prevent fires from becoming major disasters.[19] It could therefore be argued that even though the cost for FMAG declarations may have increased, FMAGs may actually save federal dollars by reducing the need for a major disaster declaration, which would decrease spending on disaster assistance programs.

Can FMAGs Still Be Issued If the DRF Balance Is Low?

The decision to issue a FMAG declaration is not contingent on the DRF balance. Similarly, FMAGs do not reduce the amount of funding available for major disasters. When the DRF balance was low in the past, FEMA used its "immediate needs funding" (INF) policy until supplemental appropriations were passed to replenish the DRF. Under INF, long-term projects (such as mitigation work) are put on hold and only activities deemed urgent are funded. It appears likely that FMAGs would fall into the category of events with an "urgent" need. Under the INF policy, FEMA also delays interagency reimbursements, and recovers funds from previous years in order to stretch its available funds.

What Are the Cost-Share Requirements for FMAGs?

As with many other Stafford Act disaster assistance grant programs (Public Assistance, Hazard Mitigation Grant assistance, Other Needs Assistance) the cost-share for FMAGs is based on a federal share of 75% of eligible expenses. The grantee (the State) and sub-grantees (local communities) assume the remaining 25% of eligible costs.[20]

Does FEMA Advance Funds to States or Reimburse States for Completed Work?

Under the FMAG process, FEMA is reimbursing grantees for eligible activities they have undertaken. The state application for grant funds must be submitted within 90 days after the FMAG is granted.

FMAG Assistance

What Types of Assistance Are Provided Under an FMAG Declaration?

FMAG assistance is similar in some basic respects to other FEMA assistance. For example, FMAGs will not replicate or displace the work of other federal agencies, nor will FEMA pay straight time salaries for public safety forces, though it will reimburse overtime expenses for the event. Other eligible expenses can include:

- Costs for equipment and supplies (less insurance proceeds);
- Costs for mobilization and demobilization;
- Costs for emergency work (evacuations and sheltering, police barricading and traffic control, arson investigation);
- Costs for pre-positioning federal, out-of-state, and international resources for up to 21 days when approved by the FEMA Regional Administrator;
- Costs for personal comfort and safety items for firefighter health and safety;
- Costs for field camps and meals in lieu of per diem; and/or
- Costs for the mitigation, management, and control of declared fires burning on co-mingled federal land, when such costs are not reimbursable by another federal agency.[21]

Is Mitigation Funding Included in an FMAG Declaration?

Currently, FMAGs do not provide any mitigation funding. However, legislation has been introduced in the 113th Congress, in both the House and Senate, to create a mitigation authority within the FMAG program. Similar language was also included in the initial Senate appropriations bill for FY2015.[22]

What Is the Time Frame from Application to Receipt of Funds?

After an FMAG has been issued for a fire, the state submits a grant application package. The package must be submitted within nine months after the FMAG is declared. That time frame permits the state to gather all information and supporting data on potentially eligible spending to include in their grant application package. The package must also stipulate that the fire cost threshold was met.[23] Following submission of the grant application FEMA has 45 days to approve or deny the application.[24]

Interaction with Other Federal Agencies

How Are FMAGs Different from Other Types of Federal Fire Assistance?

FEMA assistance through FMAGs is a direct relationship with the states to assist the state in fighting the fire on state lands. FMAGs are employed so a disaster declaration may not be necessary. The Forest Service and other federal agencies do provide other types of assistance related to wildfire management, such as post-fire recovery assistance, or assistance planning and mitigating the potential risk from future wildfires. Most of these programs provide financial and technical assistance to state partners.[25] In addition, other USDA agencies administer various other programs to provide disaster recovery assistance to nonfederal forest landowners, including the Emergency Forest Restoration Program and the Emergency Watershed Program.[26]

Can FMAG Assistance Be Provided in Conjunction with Assistance from the Forest Service, or Is It Considered a Duplication of Benefits?

This depends on the type of assistance being provided by the Forest Service. FMAG assistance is not generally available in conjunction with emergency suppression assistance from the Forest Service, or any other federal agency engaged in suppression operations. FMAGs provide assistance for suppression operations on nonfederal lands; whereas suppression operations on federal lands are the responsibility of the federal agency with jurisdiction. Limited exceptions may occur for declared fires on lands which the ownership is comingled federal and nonfederal, and the costs incurred by the eligible entity are not entitled to any other type of federal reimbursement.[27] However, FMAGs may be provided in conjunction with other Forest Service assistance programs, such as any technical and financial assistance provided through the agency's state and volunteer fire assistance programs or state and private forestry office.

FMAG and other federal assistance may potentially occur in conjunction when there is a cooperative agreement between federal, state, and other governmental or tribal partners to coordinate emergency wildfire protection and response activities.[28] The cooperative agreement often delineates different geographic areas where the state government is responsible for initial suppression operations, regardless of land ownership, and vice versa, where the federal government may be responsible for providing suppression operations in lands under nonfederal ownership. The cooperative agreements (sometimes referred to as "fire compacts") specify how costs are to be apportioned amongst the partners, including provisions allowing for reimbursement, in accordance with applicable federal and state statutes. In the circumstance where a state (or other eligible entity) conducted suppression operations on federal land and the costs were not reimbursable, an FMAG may potentially be applied for and used to cover eligible costs.

Do FMAGs Assist with Fires on Federal Lands?

No, most fires that begin on federal land are the responsibility of the federal agency that owns or manages the land, and are not eligible to receive FMAG assistance.[29] There are some exceptions, however. For example, FMAGs may be available to assist with declared fires that occur on co-mingled federal and nonfederal land, if the state has a responsibility for suppression activities under a cooperative agreement with the applicable federal agency, and those costs are not reimbursable under another federal statute.[30]

End Notes

[1] 42 U.S.C. §5121 et seq. For further analysis on the Stafford Act see CRS Report RL33053, *Federal Stafford Act Disaster Assistance: Presidential Declarations, Eligible Activities, and Funding*, by Francis X. McCarthy.

[2] 44 C.F.R. §204.24. There are ten FEMA regions in the United States and its territories. Each region is headed by a FEMA Administrator who oversees all policy, managerial, resource, and administrative actions that affect the region. The FEMA Administrator is also responsible for ensuring that policies, programs, and administrative and management guidance are implemented in a manner consistent with FEMA's overall goals.

[3] P.L. 93-288, codified at 42 U.S.C. §5187(a).

[4] 44 C.F.R §152.

[5] 44 C.F.R. §206.392. See also Federal Emergency Management Agency, *Fire Management Assistance Grant Program Guide*, FEMA P-954, February 2014, p. 10, at http://www.fema.gov/media-library-data/1394820975537-a279bff2a4a300676b870154acec922b/FMAG%20Guide%20Feb%202014_508.pdf.

[6] Section 1110 of the Sandy Recovery Improvement Act of 2013 amends Sections 401 and 501 of the Stafford Act which contain the procedures for requesting types of disaster declarations. Previously, tribal groups were treated as local governments and thus not permitted to directly request declarations from the federal government. As with local governments, the

tribes were dependent on a request being made by the governor of the state where their territory is located.

[7] DHS/FEMA, *Tribal Declarations Pilot Guidance, First Draft,* April 2014. http://www.fema.gov/media-library-data/1396364830332-2fb965ca421ca78de333ed 02c0 b5fa47/First+Draft+Tribal+Declarations+Pilot+Guidance.pdf. Under this phase guidance is developed eventually leading into the regulatory process.

[8] For additional information, see http://www.nifc.gov/index.html.

[9] Information provided by FEMA's Office of Congressional Affairs, April 24, 2014.

[10] Declaration forms can be located at http://www.fema.gov/media-library /assets/documents/ 26237?id=5789.

[11] The formula is [(population) x 0.05 x $1.39]. The dollar amount is periodically adjusted for inflation by FEMA.

[12] For more information on emergency and major disaster declarations see CRS Report RL34146, *FEMA's Disaster Declaration Process: A Primer*, by Francis X. McCarthy.

[13] Daniel Richardson, Research Assistant, Government and Finance Division, coauthored this section.

[14] For more information on the DRF see CRS Report R43537, *FEMA's Disaster Relief Fund: Overview and Selected Issues*, by Bruce R. Lindsay.

[15] For more information on the BCA and disaster relief see CRS Report R42352, *An Examination of Federal Disaster Relief Under the Budget Control Act*, by Bruce R. Lindsay, William L. Painter, and Francis X. McCarthy.

[16] Under the Stafford Act, a major disaster declaration triggers assistance to state and local governments and certain nonprofit organizations for the repair or restoration of public infrastructure, such as roads and buildings. A major disaster declaration may also include additional programs beyond temporary housing such as disaster unemployment assistance and crisis counseling, and other recovery programs, such as community disaster loans. Emergency declarations authorize activities that can help states and communities carry out essential services and activities that might reduce the threat of future damage. Emergency declarations, however, do not provide assistance for repairs and replacement of public infrastructure or nonprofit facilities. Surge funding includes funding for the execution of pre-declaration, preliminary damage assessments and for activities necessary to prepare for, and be positioned for, an incident for which a declaration under the Stafford act is likely imminent.

[17] Fire suppression obligations includes funding for infrastructure, mission assignments, and associated FEMA administration costs.

[18] Surge activities represent expenditures by FEMA prior to a Presidential declaration which may then be charged to the DRF. This is spending engaged in by FEMA when there is a reasonable expectation that a declaration is imminent.

[19] A major disaster declaration is made as a result of the disaster or catastrophic event and constitutes a broader authority that helps states and local communities, as well as families and individuals, recover from the damage caused by the event.

[20] While most FEMA assistance is cost-shared on a 75/25 basis a significant exception is the temporary housing program which is 100% federal funding.

[21] DHS/FEMA, Fire Management Assistance Grant, *Fact Sheet,* March, 2014.

[22] S. 1396; S. 1428, PREPARE Act of 2013; and H.R. 3333, Wildfire Prevention Act of 2013.

[23] Other agreements that must be in place include an approved State Administrative Plan, a FEMA-State Agreement for the incident and that the state has an approved State Mitigation Plan. While these are important steps they are also part of the ongoing, operational

relationship between the state and the FEMA regional office and have likely been accomplished or can be accomplished or revised during the nine-month time frame.

[24] DHS/FEMA, *Fire Management Assistance Grant Program,* Intergovernmental Webinar Series, June 26, 2014.

[25] For more information, see CRS Report RL31065, *Forestry Assistance Programs*, by Katie Hoover.

[26] For more information see CRS Report R42854, *Emergency Assistance for Agricultural Land Rehabilitation*, by Megan Stubbs.

[27] 44 C.F.R. §204.42(i).

[28] For an example of a federal-state fire compact, see the California Master Cooperative Wildland Fire Management and Stafford Act Response Agreement, available at http://www.fs.usda.gov/detail/r5/fire-aviation/ ?cid=stelprdb537432. For an example of a federal, state, and international fire compact, see the Northeast Forest Fire Protection Compact, available at http://www.nffpc.org.

[29] 44 C.F.R. §204.43(e).

[30] 44 C.F.R. §204.42(i).

INDEX

D

E

F

G

H

I

J

L

M

N